Famous Problems
of Geometry
and How to Solve Them

BENJAMIN BOLD

DOVER PUBLICATIONS, INC.
New York

Published in Canada by General Publishing Company, Ltd.,
30 Lesmill Road, Don Mills, Toronto, Ontario.
Published in the United Kingdom by Constable and Company,
Ltd., 10 Orange Street, London WC2H 7EG.

This Dover edition, first published in 1982, is an unabridged
and slightly corrected republication of the work originally published in 1969 by Van Nostrand Reinhold Company, N.Y., under
the title *Famous Problems of Mathematics: A History of Constructions with Straight Edge and Compass.*

Manufactured in the United States of America
Dover Publications, Inc.
180 Varick Street
New York, N.Y. 10014

Library of Congress Cataloging in Publication Data

Bold, Benjamin.
Famous problems of geometry and how to solve them.

Reprint. Originally published: Famous problems of mathematics. New York : Van Nostrand Reinhold, 1969.
1. Geometry—Problems, Famous. I. Title.
QA466.B64 1982   516.2'04   81-17374
ISBN 0-486-24297-8   AACR2

To My Wife, CLAIRE,

Whose Assistance and Encouragement
Made this Book Possible

Contents

vii

Foreword

In June of 1963 a symposium on Mathematics and the Social Sciences was sponsored by the American Academy of Political and Social Sciences. One of the contributions was by Oscar Morgenstern, who, together with John Von Neumann, had written the book "The Theory of Games and Economic Behavior." This book stimulated the application of mathematics to the solution of problems in economics, and led to the development of the mathematical "Theory of Games." Dr. Morgenstern's contribution to the Symposium was called the "Limits of the Uses of Mathematics in Economics." I shall quote the first paragraph of this article.

> Although some of the profoundest insights the human mind has achieved are best stated in negative form, it is exceedingly dangerous to discuss limits in a categorical manner. Such insights are that there can be no perpetuum mobile, that the speed of light cannot be exceeded, that the circle cannot be squared using ruler and compasses only, that similarly an angle cannot be trisected, and so on.

Each one of these statements is the culmination of great intellectual effort. All are based on centuries of work and either on massive empirical evidence or on the development of new mathematics or both. Though stated negatively, these and other discoveries are positive achievements and great contributions to human knowledge. All involve mathematical reasoning; some are, indeed, in the field of pure mathematics, which abounds in statements of prohibitions and impossibilities."

The above quotation states clearly and forcefully the purpose of this book. Why does mathematics abound in "statements of prohibitions and impossibilities?" Why are the solutions of such problems as "squaring the circle" and "trisecting an angle" considered to be "profound insights" and "great contributions to human knowledge?" Why were centuries of "great intellectual effort" required to solve such seemingly simple problems? And, finally, what new mathematics had to be developed to resolve these problems? I hope you will find the answers to these questions as you read this book.

The outstanding achievement of the Greek mathematicians was the development of a postulational system. Despite the flaws and defects of euclidean geometry as conceived by the ancient Greeks, their work serves as a model that is followed even up to the present day.

In a postulational system one starts with a set of unproved statements (postulates) and deduces (by means of logic) other statements (theorems). Two of the postulates of euclidean plane geometry are:

1) given any two distinct points, there exists a unique line through the two points.

2) given a point and a length, a circle can be constructed with the given point as center and the given length as radius.

These two postulates form the basis for euclidean constructions (constructions using only an unmarked straight edge and compasses). With these two instruments the Greek mathematicians were able to perform many constructions; but they also were unsuccessful in many instances. Thus, they were able to bisect any given angle, but were unable to trisect a general angle. They were able to construct a square equal in area to twice a given square, but were unable to "duplicate a cube." They were able to construct a square equal in area to a given polygon, but were unable to "square a circle." They were able to construct regular polygons of 3, 4, 5, 6, 8 and 10 sides, but were unable to construct regular polygons of 7 or 9 sides. Before the end of the 19th century, mathematicians had supplied answers to all of these problems of antiquity. The purpose of this book is to show how these problems were eventually solved.

Why were the Greek mathematicians unable to solve these problems? Why was there a lapse of about two thousand years before solutions to these problems were found?

The mathematical efforts of the Greeks were along geometric lines. The concentration on geometry, and the resulting neglect of algebra, was due to the following situation:

The Pythagorean theorem tells us that, if the length of a side of a square is one unit, the length of the diagonal is $\sqrt{2}$ units. What kind of number is $\sqrt{2}$? The Greek mathematicians, up to this point, were able to express all their results in terms of integers. Fractions, or rational numbers, are ordered pairs of integers—i.e., numbers of the form a/b, where a and b are integers, $b \neq 0$. No matter how they tried, the Greeks were unable to express $\sqrt{2}$ in terms of integers. As you already know, it can be proven that $\sqrt{2}$ is irrational, and it was not until the 19th century that a satisfactory theory of irrationals was developed.

Because of the lack of such a theory, the course of Greek

mathematics took a geometric turn. Thus, when the Greeks wished to expand $(a + b)^2$, they proceeded geometrically as follows:

$$(a + b)^2 = a^2 + 2ab + b^2$$

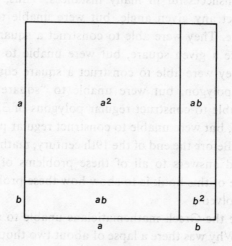

As we shall show later the solution of the construction problems involves well developed algebraic techniques, and in fact, it was not until algebra and analytic geometry were developed in the 17th century by Vieta, Descartes, Fermat, and others, that procedures were obtained that could be used in a successful attack on the construction problems.

Famous Problems
of Geometry
and How to Solve Them

CHAPTER I

Achievement of the Ancient Greeks

USING GEOMETRIC THEOREMS, the Greek mathematicians were able to construct any desired geometric element that could be derived by a finite number of rational operations and extractions of real square roots from the given elements. To illustrate: Suppose we are given the elements a, b, and the unit element. The Greeks could construct $a + b$, $a - b$, $a \cdot b$, a/b, a^2, and $\sqrt{a}$.

PROBLEM SET I–A

Construct $a + b$ and $a - b$, using the given line segments. The following diagram shows how to construct ab. If EG is

1

constructed parallel to DF, then $x = ab$.

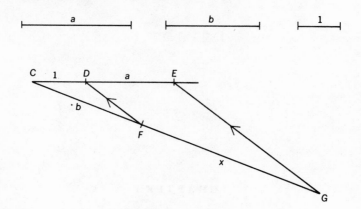

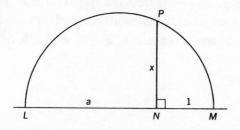

PROBLEM SET I–B

1. Prove that $x = ab$.
2. Using a similar procedure construct a^2; a/b; a^2/b
   The diagram below shows how to construct $\sqrt{a}$.

A semicircle is constructed on LM as a diameter. NP is perpendicular to LM (P is the intersection of the perpendicular and the semicircle). Then $x = \sqrt{a}$.

PROBLEM SET I–C

1. Using the diagram above, prove $x = \sqrt{a}$.
2. Using a similar procedure construct

$$\sqrt{ab}; \quad \sqrt[4]{a}; \quad \sqrt[8]{a}$$

3. Using the Pythagorean theorem construct line segments equal to

$$\sqrt{2}; \quad \sqrt{3}; \quad \sqrt{5}; \quad \sqrt{17}$$

Using these constructions, the Greeks were able to construct the roots of a linear or quadratic equation if the numbers representing the coefficients were the lengths of given line segments.

PROBLEM SET I–D

Construct the root of $ax + b = c$, where a, b, c are given line segments.

To construct the roots of the quadratic equation $x^2 - ax + b = 0$ $(a^2 > 4b)$, one can proceed as follows: Construct a circle whose diameter BD joins the points B(0, 1) and D(a, b). Then the abscissas of G and F (the points where the circle intersects the X-axis) will be the roots of the quadratic equation.

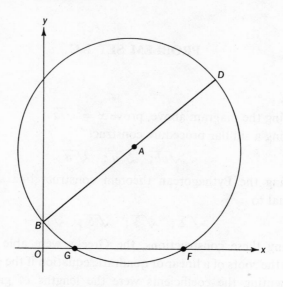

PROBLEM SET I–E

1. Why do we use the restriction $a^2 > 4b$?
2. Prove that the abscissas of G and F are the roots of $x^2 - ax + b = 0$. Hint: Show that the equation of the circle is

$$\left(x - \frac{a}{2}\right)^2 + \left(y - \frac{b+1}{2}\right)^2 = \frac{a^2}{4} + \frac{(b-1)^2}{4}$$

As pointed out in the Introduction, the Greeks (using the basic constructions outlined in this section) achieved considerable success in construction problems. Yet they left a number of unsolved problems for future generations of mathematicians to struggle with. The remainder of this book will be devoted to a brief history of attempts to solve these problems and their final solution in the nineteenth century.

PROBLEM SET I–F

Construct the positive root of $x^2 + x - 1 = 0$, given a unit length.

CHAPTER II

An Analytic Criterion for Constructibility

To ANSWER THE QUESTION "Which constructions are possible with unmarked straight edge and compasses?" it is necessary to establish an analytic criterion for constructibility. Every construction problem presents certain given elements a, b, c $\cdots$ and requires us to find certain other elements x, y, z $\cdots$. The conditions of the problem enable us to set up one or more equations whose coefficients will be numbers representing the given elements a, b, c $\cdots$. The solutions of the equations will permit us to express the unknown elements in terms of the given elements. To take a simple example, suppose we wish to construct a square equal in area to twice a given square whose side is a. We express the problem analytically by the equation

$$x^2 = 2a^2$$

Other problems, of course, will lead to equations of higher degree.

We have already seen how the roots of a linear or quadratic equation can be constructed. We shall now investigate the possibility of constructing the roots of an equation of degree greater than 2, and we shall be especially interested in the roots of cubic equations.

First, we know that it is possible to construct any geometric element if that element can be derived from the given elements by a finite number of rational operations (addition, subtraction, multiplication and division), and the extraction of real square roots. Now let us consider the converse situation. If a construction is possible, what is the relation between the required elements and the given elements? It is easy to see that only those constructions are possible for which the numbers which define the desired elements can be obtained from the given elements by a finite number of rational operations and the extractions of real square roots.

Any construction consists of a sequence of steps, and each step is one of the following:

1   drawing a straight line between two points.
2   constructing a circle with a given center and given radius.
3   finding the points of intersection of two straight lines, two circles, or a straight line and a circle.

Let us assume that we are given a set of coordinate axes and a unit length and that all the given elements can be represented by rational numbers. We know that the sum, difference, product, and quotient (division by 0 is always excluded) of two rational numbers is a rational number. The rational numbers are said to form a closed set with respect to the four fundamental operations. Any set of numbers closed with respect to these four operations is called a field. Let us represent the field of rational numbers by F_0.

If we are given the coordinates of two points

$$P_1(x_1, y_1) \quad \text{and} \quad P_2(x_2, y_2)$$

then the equation of the line thru $P_1 P_2$ is

$$(y_2 - y_1)x + (x_1 - x_2)y + (x_2y_1 - x_1y_2) = 0$$

or

$$ax + by + c = 0$$

where

$$a = y_2 - y_1; \, b = x_1 - x_2; \, \text{and} \, c = x_2y_1 - x_1y_2$$

Note that x_1, x_2, y_1, and y_2 are rational numbers by definition. Then a, b, and c are also rational numbers.

The equation of a circle, whose center is (h, k) and whose radius is r, is

$$x^2 + y^2 - 2hx - 2ky + h^2 + k^2 - r^2 = 0$$

or

$$x^2 + y^2 + dx + ey + f = 0$$

where $d = -2h$, $e = -2k$ and $f = h^2 + k^2 - r^2$. Again, d, e, and f are rational numbers. Finding the coordinates of the point of intersection of two straight lines involves only rational operations performed on the coefficients of the variables, whereas finding the coordinates of the intersection of a straight line and a circle, or of two circles, would involve, in addition to the rational operations, only the extraction of square roots. To summarize, we can state that a proposed construction with unmarked straight edge and compasses is possible *if and only if* the numbers which define the desired elements can be derived from the given elements by a finite number of rational operations and the extractions of square roots.

PROBLEM SET II–A

Find the coordinates of the points of intersection of

(a) $ax + by + c = 0$ and $a'x + b'y + c' = 0$

(b) $ax + by + c = 0$ and $x^2 + y^2 + dx + ey + f = 0$

(c) $x^2 + y^2 + dx + ey + f = 0$ and

$$x^2 + y^2 + d'x + e'y + f' = 0$$

All rational numbers, i.e., all numbers in F_0, can be constructed if we are given a unit length. Furthermore, if k is a fixed number in F_0 we can construct $\sqrt{k}$ and $a + b\sqrt{k}$, where a and b are any numbers in F_0. If $\sqrt{k}$ is not in F_0, then we can prove that all numbers of the form $a + b\sqrt{k}$ (a,b,k in F_0) form a new field F_1, which has F_0 as a subfield. For example, let $k = 3$, then $a + b\sqrt{3}$ form a field F_1 which contains all rational numbers as a subfield, namely all those numbers for which $b = 0$.

To prove that all numbers $a + b\sqrt{k}$ (a, b, k in F_0, $\sqrt{k}$ not in F_0) form a field we observe that

1. $(a + b\sqrt{k}) + (c + d\sqrt{k}) = (a + c) + (b + d)\sqrt{k}$
$$= e + f\sqrt{k}$$

2. $(a + b\sqrt{k}) - (c + d\sqrt{k}) = (a - c) + (b - d)\sqrt{k}$
$$= m + n\sqrt{k}$$

3. $(a + b\sqrt{k}) \times (c + d\sqrt{k}) = (ac + bdk) + (ad + bc)\sqrt{k}$
$$= r + s\sqrt{k}$$

4. $\dfrac{a + b\sqrt{k}}{c + d\sqrt{k}} \cdot \dfrac{c - d\sqrt{k}}{c - d\sqrt{k}} = \dfrac{(ac - bdk) + (bc - ad)\sqrt{k}}{c^2 - d^2k}$

$$= \frac{ac - bdk}{c^2 - d^2k} + \frac{(bc - ad)\sqrt{k}}{c^2 - d^2k}$$

$$= u + v\sqrt{k}$$

Since a, b, c, d, k are numbers in the field F_0, the sum, product, difference, and quotient of any two of these rational numbers is a rational number. Therefore e, f, m, n, r, s, u, and v are rational numbers, which proves that all numbers of the form $a + b\sqrt{k}$ form a field F_1, which contains F_0 as a subfield. There is one detail which requires clarification. When $(a + b\sqrt{k})$ was divided by $(c + d\sqrt{k})$ we obtained $u + v\sqrt{k}$

$$u = \frac{ac - bdk}{c^2 - d^2k}$$

and

$$v = \frac{bc - ad}{c^2 - d^2k}$$

Thus u and v are rational numbers if and only if

$$c^2 - d^2k \neq 0$$

We have assumed that $\sqrt{k}$ is not in F_0. Therefore, $k \neq 0$. Also when we divide we assume the divisor is not 0. Therefore c and d are not both 0, although either c or d may be 0. We now have to show that $c^2 - d^2k \neq 0$. If $c^2 - d^2k = 0$, then $c^2 = d^2k$, $k = c^2/d^2$ and $\sqrt{k} = \pm c/d$. Thus $\sqrt{k}$ would be in F_0 contrary to the hypothesis. This completes the proof that all numbers of the form $a + b\sqrt{k}$ form a field, F_1.

Next we can construct all numbers of the form $a_1 + b_1\sqrt{k_1}$ where a_1 and b_1 are any numbers in F_1, k_1 is a fixed number in F_1, and the $\sqrt{k_1}$ is not in F_1. Such a number would be $\sqrt{5} + \sqrt{2}\sqrt[4]{3}$, where $a_1 = \sqrt{5}$, $b_1 = \sqrt{2}$, $k_1 = \sqrt{3}$ are in F_1 and $\sqrt{k_1} = \sqrt[4]{3}$ is not in F_1. Another example would be $5 + 2\sqrt[4]{3}$. Again we can prove by a process completely analogous to that used above that all numbers of the form $a_1 + b_1\sqrt{k_1}$ form a field F_2, which has F_1 as a subfield (which contains numbers of the form $a_1 + b_1\sqrt{k_1}$, where $b_1 = 0$).

The process can be continued indefinitely until we reach a field F_n, n, a positive integer. We now have a sequence of fields $F_0, F_1, F_2, \cdots F_n$ with the following properties:

1. F_0 is the field of rational numbers.

F_1 is the field obtained by adjoining $\sqrt{k}$ to F_0, where k is a fixed number in F_0 such that $\sqrt{k}$ is not in F_0; F_1 contains all numbers of the form $a + b\sqrt{k}$, a, b, k in F_0, $\sqrt{k}$ not in F_0.

F_2 is the field obtained by adjoining $\sqrt{k_1}$ to F_1, where k_1 is a fixed number in F_1 such that $\sqrt{k_1}$ is not in F_1. The field F_2 contains all numbers of the form $a_1 + b_1\sqrt{k_1}$; a_1, b_1, k_1 in F_1, $\sqrt{k_1}$ not in F_1.

F_n is the field obtained by adjoining $\sqrt{k_{n-1}}$ to F_{n-1}, where k_{n-1} is a fixed number in F_{n-1} such that $\sqrt{k_{n-1}}$ is not in F_{n-1}. F_n contains all numbers of the form

$$a_{n-1} + b_{n-1}\sqrt{k_{n-1}}; a_{n-1}, b_{n-1}, k_{n-1} \quad \text{in} \quad F_{n-1}$$

but $\sqrt{k_{n-1}}$ is not in F_{n-1}. A number in F_n may be exceedingly complicated, consisting as it does of n square roots, one over the other. An example of a number in F_4 would be

$$\sqrt{2} - \sqrt{\sqrt{3 + \sqrt{7 + \sqrt{5}}}}$$

To obtain the number

$$\sqrt{\sqrt{3 + \sqrt{7 + \sqrt{5}}}}$$

we can start with $k_0 = 5$, then $k_1 = 7 + \sqrt{k_0}$, $k_2 = 3 + \sqrt{k_1}$, $k_3 = \sqrt{k_2}$, and $k_4 = \sqrt{k_3}$. This number is, therefore, in F_4.

2. F_0 is a subfield of F_1; F_1 is a subfield of F_2; F_{n-1} is a subfield of F_n.

3. Every number in $F_0, F_1, F_2, \cdots, F_n$ is constructible, since a number in any one of these fields can be obtained from the unit element by a finite number of rational operations and extractions of square roots.

4. Conversely any constructible number can be found in one of the fields $F_0, F_1, F_2, \cdots, F_n$, since we have already shown that only those constructions are possible for which the numbers which define the desired elements can be obtained from the given elements (which we assumed to be represented

by rational numbers) by a finite number of rational operations and extractions of square roots.

As a summary we can state the following theorem—All the numbers in the fields F_0, F_1, $\cdots$, F_n are constructible, and conversely, any constructible number must be in one of the fields F_0, F_1, $\cdots$, F_n. Thus, $\sqrt{3 + \sqrt[8]{2}}$ is constructible, since $\sqrt{3 + \sqrt[8]{2}}$ is in F_4; whereas $\sqrt{3 + \sqrt[6]{2}}$ is not constructible, since $\sqrt{3 + \sqrt[6]{2}}$ is not in any one of the fields F_0, F_1, $\cdots$, F_n.

PROBLEM SET II–B

1. Show that $7/(5 - \sqrt{2})$ is in F_1 by expressing the number in the form $a + b\sqrt{k}$, where a, b, k are in F_0.
2. Show that $5/(2 - \sqrt[4]{3})$ is in F_2 by expressing the number in the form $a_1 + b_1\sqrt{k_1}$, where a_1, b_1, k_1 are in F_1. Hint: To rationalize the denominator, use the identity

$$s^4 - t^4 = (s - t)(s + t)(s^2 + t^2), s = 2, t = \sqrt[4]{3}$$

We are now in a position to determine when the roots of a cubic equation are constructible, by proving the following theorem: If a cubic equation with rational coefficients has no rational root, then none of its roots is constructible. Any cubic equation with rational coefficients is said to be reducible in the field of rational numbers if it has at least one rational root. If the equation has no rational root, it is said to be irreducible in the field of rational numbers. Thus, we wish to show that no root of an irreducible cubic equation can be constructed. We can represent the cubic equation by

$$x^3 + px^2 + qx + r = 0$$

where p, q, r are in F_0. By hypothesis, the equation is irreducible, and consequently has no rational roots. Let us assume that one of the roots, x_1, is constructible. Then x_1 is a number in some field F_n, where n is an integer > 0. x_1 cannot belong to F_0, since the equation is irreducible. Let m be the smallest integer for which x_1 belongs to F_m, i.e., x does not belong to F_{m-1} $(m > 1)$. If the equation has any other constructible root, we assume that that root belongs to F_r, where $r \geq m$. Then x_1 is of the form

$$a_{m-1} + b_{m-1}\sqrt{k_{m-1}}$$

To simplify our notation, let

$$a = a_{m-1}, b = b_{m-1}, k = k_{m-1}$$

In other words, a, b, k belong to F_{m-1}, and $x_1 = a + b\sqrt{k}$ belongs to F_m but not to F_{m-1}.

We now show that if $a + b\sqrt{k}$ is a root of the cubic equation

$$x^3 + px^2 + qx + r = 0$$

then $a - b\sqrt{k}$ must also be a root. If $a + b\sqrt{k}$ is a root of the equation, then

$$(a + b\sqrt{k})^3 + p(a + b\sqrt{k})^2 + q(a + b\sqrt{k}) + r = 0$$

Simplifying, we obtain

$$a^3 + 3a^2b\sqrt{k} + 3ab^2k + b^3k\sqrt{k} + pa^2$$
$$+ 2pab\sqrt{k} + pb^2k + qa + qb\sqrt{k} + r = 0$$

or $s + t\sqrt{k} = 0$, where

$$s = a^3 + 3ab^2k + pa^2 + pb^2k + qa$$
$$+ r \quad \text{and} \quad t = 3a^2b + b^3k + 2pab + qb$$

Thus, either $\sqrt{k} = -s/t$ or $s = 0$ and $t = 0$. However, t and s are numbers in F_{m-1}. Therefore, if $\sqrt{k} = -s/t$, $\sqrt{k}$ would also be a number in F_{m-1}, contrary to the hypothesis that m is the least integer for which x_1 belongs to F_m. Therefore, $s = 0$ and $t = 0$.

If we now substitute $a - b\sqrt{k}$ for x in the polynomial

$$x^3 + px^2 + qx + r = 0$$

we obtain the expression $s' + t'\sqrt{k}$, where $s' = a^3 + 3ab^2k + pa^2 + pb^2k + qa + r$ and $t' = -(3a^2b + b^3k + 2pab + qb)$. Thus $s' = s = 0$, and $t' = -t = 0$. Therefore, $s' + t'\sqrt{k} = 0$ and $a - b\sqrt{k}$ is a root of the cubic equation.

We now know that $x_1 = a + b\sqrt{k}$ and $x_2 = a - b\sqrt{k}$. To find the third root x_3, we observe that the cubic equation can be written in the form

$$(x - x_1)(x - x_2)(x - x_3) = 0$$
$$x^3 - (x_1 + x_2 + x_3)x^2 + (x_1x_2 + x_1x_3 + x_2x_3)x - x_1x_2x_3 = 0.$$

Therefore

$$x_1 + x_2 + x_3 = -p$$

or, since $x_1 + x_2 = 2a$, $x_3 = -2a - p$, which means that one of the roots, x_3, is in the field F_{m-1}, contrary to the hypothesis that m is the least integer such that F_m contains a root of the cubic equation.

We can, therefore, conclude that if a cubic equation with rational coefficients has a constructible root, it also has a rational root. If we represent the rational root by x_r, we can write the cubic equation as

$$(x - x_r)(x^2 + p_1x + q_1) = 0.$$

Consequently, the other two roots are roots of a quadratic equation and are also constructible. Conversely, if a cubic equation with rational coefficients has a rational root, we can write the equation in the form

$$(x - x_r)(x^2 + p_1x + q_1) = 0$$

where p_1 and q_1 are rational. Thus the roots of this equation are constructible. To summarize, we can state the following theorem:

The roots of a cubic equation with rational coefficients are constructible if and only if the equation has a rational root. If the equation is irreducible, then none of its roots can be constructed with straight edge and compasses.

This property of cubic equations is basic in solving most of the construction problems the Greeks were unable to dispose of. As later chapters will show, this theorem can be used to solve the following problems—duplicating a cube, trisecting an arbitrary angle, constructing a regular polygon of 7 sides and 9 sides. The problem of squaring a circle requires results of a different nature, as the development will show. Before we undertake the assault on these problems, there will be a brief interlude on Complex Numbers.

Finally, it is interesting to observe that the theorem on cubic equations is a special case of a more general theorem whose proof is beyond the scope of this book. We first define the term *irreducible* in a more general context: Let $p(x) = a_0 x^n + a_1 x^{n-1} + \ldots + a_n = 0$ be a polynomial equation of degree n, where n represents a positive integer and a_i represents a rational number. Then $p(x) = 0$ is reducible in the field of rational numbers if $p(x)$ can be factored into polynomials of *lower* degree with coefficients in the field of rational numbers. If $p(x)$ cannot be factored in that manner then $p(x)$ is said to be irreducible. It can now be proven that a geometric element is constructible if, and only if, the number representing the element is the root of an irreducible polynomial equation (with rational coefficients) of degree 2^k, where k represents a non-negative integer.

PROBLEM SET II–C

1. Show that $\sqrt{3 + \sqrt[4]{2}}$ is the root of an irreducible equation of degree 2^3.
2. Construct $\sqrt{3 + \sqrt[4]{2}}$.
3. Which one of the cubic equations has a rational root
   (a) $x^3 - 1 = 0$
   (b) $x^3 - 2 = 0$
4. Find the three roots of the reducible cubic equation.
5. Given a cubic equation $x^3 + ax^2 + bx + c = 0$, where a, b, and c are rational numbers, and whose roots r_1, r_2, and r_3 are positive real numbers; if r_1 is rational, show that all the roots of the equation are constructible by showing that the roots are numbers in F_0 or F_1. As an illustration, construct the roots of $x^3 - 7x^2 + 14x - 6 = 0$ given a unit length.
6. If $2 + \sqrt{3}$ is a root of the cubic equation $x^3 + ax^2 + bx + c = 0$, where a, b, and c are rational numbers, show that one of the roots of the equation is rational.

CHAPTER III

Complex Numbers

YOU MAY WONDER what complex numbers have to do with constructing geometric lines or figures. But as you progress you will observe that complex numbers are most useful in both algebraic and geometric problems involving real numbers.

You will recall that finding the diagonal of a square of unit side led to the equation $x^2 = 2$, and that the solution of this equation proved to be a real obstacle to the progress of Greek mathematics. The Greeks could obtain the rational numbers by forming ordered pairs of integers (a/b, $b \neq 0$). However, there was no such simple procedure for proceeding from the rationals to the reals (including irrational numbers).

Another roadblock in the history of mathematics resulted from the equation $x^2 + 1 = 0$. You should be able to prove that there is no real number x whose square is -1. (The square of a positive or negative number is positive, and the square of 0 is 0.) There are two alternatives: either we must assume that the simple equation $x^2 + 1 = 0$ has no solution,

19

or we must extend the real number system in such a way that the new number field will contain a solution of $x^2 + 1 = 0$. The latter alternative is, of course, the more desirable one, and the new number field can be obtained easily by following a procedure similar to that used in forming rational numbers from the integers, i.e., we form a complex number by using an ordered pair of real numbers with suitable definitions for equality, addition, and multiplication. We can represent a complex number in the form $a + bi$ ($i = \sqrt{-1}$). The definitions for equality, addition, and multiplication will be as follows:

1 $a + bi = c + di$ if and only if $a = c$ and $b = d$
2 $(a + bi) + (c + di) = (a + c) + (b + d)i$
3 $(a + bi) \cdot (c + di) = (ac - bd) + (ad + bc)i$

These are the definitions you would expect if we consider i to behave as any other variable such as x, with the restriction that $i^2 = -1$.

PROBLEM SET III–A

For which values of a and b will $a + bi$ be a solution of $x^2 + 1 = 0$.

It is not necessary to use i in representing a complex number. We can merely write (a, b), where a and b are real numbers.

PROBLEM SET III–B

1. Write the definitions for equality, addition, and multi-

plication, using the form (a, b) to represent a complex number.

2. If we write the additive identity for complex numbers as $(0, 0)$, and the multiplicative identity as $(1, 0)$, find the solution of $x^2 + (1, 0) = (0, 0)$ and check your answers.

You are no doubt familiar with the fact that polar coordinates (r, θ) as well as rectangular coordinates (a, b) can be used to represent a complex number. Then, in polar form, the complex number $a + bi$, or (a, b) can be represented as $r(\cos \theta + i \sin \theta)$.

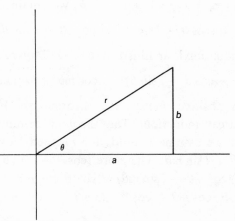

The polar form of complex numbers enables us to multiply two complex numbers very easily.

$$r_1 (\cos \theta_1 + i \sin \theta_1) \cdot r_2 (\cos \theta_2 + i \sin \theta_2)$$
$$= r_1 r_2 [\cos (\theta_1 + \theta_2) + i \sin (\theta_1 + \theta_2)]$$

i.e., when we multiply two complex numbers we multiply the moduli and add the amplitudes.

PROBLEM SET III–C

1. Prove the last statement using the trigonometric identities

$$\sin (x + y) = \sin x \cos y + \cos x \sin y$$
$$\cos (x + y) = \cos x \cos y - \sin x \sin y$$

2. Prove that

$$\frac{r_1(\cos \theta_1 + i \sin \theta_1)}{r_2(\cos \theta_2 + i \sin \theta_2)} = \frac{r_1}{r_2}[\cos (\theta_1 - \theta_2) + i \sin (\theta_1 - \theta_2)]$$

If we let $r_1 = r_2 = 1$ and $\theta_1 = \theta_2$ we obtain

$$(\cos \theta + i \sin \theta)^2 = \cos 2\theta + i \sin 2\theta$$

The last equation is an illustration of De Moivre's theorem:

$$(\cos \theta + i \sin \theta)^m = \cos m\theta + i \sin m\theta$$

For m a positive integer, the theorem can be proven by mathematical induction. The theorem certainly holds for $m = 1$. If we assume it holds for m, then $(\cos \theta + i \sin \theta)^m = \cos m\theta + i \sin m\theta$. Therefore $(\cos \theta + i \sin \theta)^{m+1} = (\cos \theta + i \sin \theta)(\cos m\theta + i \sin m\theta) = \cos \theta \cos m\theta - \sin \theta \sin m\theta + i \ (\sin \theta \ \cos m\theta + \cos \theta \ \sin m\theta) = \cos (m + 1)\theta + i \sin (m + 1)\theta$

Thus, the theorem is shown to hold for all positive integers. However, De Moivre's theorem can be established for all real or complex values of m. Let us use De Moivre's theorem to obtain a useful formula for *real numbers*.

If x is a real number, then $(\cos x + i \sin x)^2 = \cos 2x + i \sin 2x$. But $(\cos x + i \sin x)^2 = (\cos^2 x - \sin^2 x) + 2i \sin x \cos x$, and $\cos 2x + i \sin 2x = (\cos^2 x - \sin^2 x) + 2i \sin x \cos x$. Therefore, $\cos 2x = \cos^2 x - \sin^2 x$, and $\sin 2x = 2 \sin x \cos x$. Observe that we were able to obtain formulas involving only real numbers by equating the real and imaginary parts of two equal complex numbers.

PROBLEM SET III-D

Using the above procedure, prove
1. $\cos 3x = 4 \cos^3 x - 3 \cos x$
2. $\sin 3x = 3 \sin x - 4 \sin^3 x$
Hint: After equating real and imaginary parts of the complex numbers, use the relation $\sin^2 x + \cos^2 x = 1$.

Note that we shall use the first formula of Problem Set III-D when we discuss the problem of trisecting an angle.

De Moivre's theorem enables us to find the roots of unity in a very simple manner. (We shall use the roots of unity when we discuss the problem of constructing regular polygons). To find the two square roots of unity we may proceed algebraically as follows:

$$x^2 = 1 \quad \text{and} \quad x = +1 \quad \text{or} \quad x = -1$$

Using De Moivre's theorem we would proceed as follows: Let $R = \cos \theta + i \sin \theta$, where R is a square root of unity. Then $R^2 = \cos 2\theta + i \sin 2\theta = 1$, since R is a square root of unity. Therefore $2\theta = 2k\pi$, where k is any integer. However, only 2 values of k will give distinct values of R. Thus

$$R_1 = \cos \pi + i \sin \pi = -1 \ (k = 1)$$
$$R_2 = \cos 2\pi + i \sin 2\pi = 1 \ (k = 2)$$

No advantage is evident in using De Moivre's theorem in the above example. However, should we wish to find the 17 seventeenth roots of unity, then the power of De Moivre's theorem would soon become clear.

Let us try finding the three cube roots of unity. Algebraically we use the equation

$$x^3 - 1 = 0$$
$$(x - 1)(x^2 + x + 1) = 0$$
$$x_1 = 1, \; x_2 = \frac{-1 + \sqrt{-3}}{2}, \; x_3 = \frac{-1 - \sqrt{-3}}{2}$$

Observe that two of the roots are imaginary numbers.

By De Moivre's theorem we have

$$R = \cos \theta + i \sin \theta$$
$$R^3 = \cos 3\theta + i \sin 3\theta = 1$$
$$3\theta = 2k\pi, \; k = 1, 2, 3$$
$$\theta = \frac{2k\pi}{3}, \; k = 1, 2, 3$$

$$R_1 = \cos \frac{2\pi}{3} + i \sin \frac{2\pi}{3}$$

$$R_2 = \cos \frac{4\pi}{3} + i \sin \frac{4\pi}{3}$$

$$R_3 = \cos \frac{6\pi}{3} + i \sin \frac{6\pi}{3} = 1$$

PROBLEM SET III–E

Show that $R_1 = x_2$, $R_2 = x_3$, $R_3 = x_1$.

Observe that (again using De Moivre's theorem)

$$R_2 = R_1^2, \; R_3 = R_1^3$$

Therefore, we can write the three cube roots of unity as R, R^2 and R^3. Also observe that the cube roots of unity, when plotted, divide the unit circle into three equal parts and that an equilateral triangle can be formed by joining the points of division.

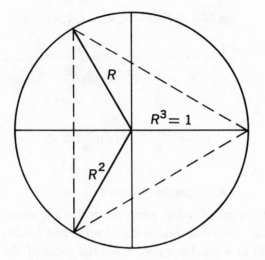

PROBLEM SET III–F

1. If we let $x_2 = (-1 + \sqrt{-3})/2 = \omega$, prove that $x_3 = \omega^2$ and $x_1 = \omega^3$. Does a similar relation hold for the two square roots of unity?

Let us now generalize the procedure and find expressions for the n roots of the equation $x^n = 1$ or $x^n - 1 = 0$. By using the formula for the sum of geometric series or simply by multiplying, one can show that

$$\frac{x^n - 1}{x - 1} = x^{n-1} + x^{n-2} + \cdots + x + 1 \quad (x \neq 1)$$

If we let $R(\neq 1)$ be an n nth root of unity, then all the roots can be expressed as $R, R^2, R^3, \cdots, R^{n-1}, R^n = 1$, since an nth root of unity can be written in the form

$$\cos \frac{2k\pi}{n} + i \sin \frac{2k\pi}{n}, \; k = 1, 2, \cdots, n$$

Thus for $k = 1$,

$$R_1 = \frac{\cos 2\pi}{n} + i \sin \frac{2\pi}{n}$$

for $k = 2$,

$$R_2 = \frac{\cos 4\pi}{n} + i \sin \frac{4\pi}{n} = R_1^2$$

for $k = n$,

$$R_n = \cos 2\pi + i \sin 2\pi = R^n = 1$$

The n nth roots of unity, when plotted in the complex plane, divide the unit circle into n equal arcs, and joining the arcs will result in a regular n-gon. Since the roots of the equation $x^{n-1} + n^{n-2} + \cdots + x + 1 = 0$ are the complex nth roots of unity and together with the root $x = 1$, divide the circle into equal parts, the equation is called the "cyclotomic" (circle dividing) equation. The numbers $R, R^2, \cdots, R^n = 1$ form a multiplicative group since they satisfy the following four conditions:

1 closure $R^a \cdot R^b = R^{a+b} = R^c$, where a, b, c are integers $\leq n$
2 associativity $R^a(R^b \cdot R^c) = (R^a \cdot R^b) \cdot R^c = R^{a+b+c}$
3 identity element is R^n since $R^a \cdot R^n = R^a$
4 inverse element for R^a is R^{n-a}

Also observe that the inverse of $R \, (\cos 2\pi/n + i \sin 2\pi/n)$ is R^{-1} which can be written in the form $\cos (2\pi/n) - i \sin (2\pi/n)$.

PROBLEM SET III–G

1. Express the seven 7th roots of unity $(R, R^2, \cdots, R^7 = 1)$ in polar form.
2. The seven 7th roots of unity form a group. Show that
   (a) $R^3 \cdot R^6$ is a member of the group.
   (b) The inverse of R^5 is a member of the group.
   (c) $R^2 + (1/R^2) = 2 \cos 4\pi/7$
3. By using the formula $S = (ar^n - a)/(r - 1)$, show that
   $$1 + R + R^2 + \cdots + R^{n-1} = \frac{R^n - 1}{R - 1}$$

CHAPTER IV

The Delian Problem

THE PROBLEM OF CONSTRUCTING a cube whose volume shall be twice that of a given cube is known as the Delian problem. D. E. Smith in his History of Mathematics relates the following story with reference to this problem: "... the Athenians appealed to the oracle at Delos to know how to stay the plague which visited their city in 430 B.C. It is said that the oracle replied that they must double in size the altar of Apollo. This altar being a cube, the problem was that of its duplication."

The problem of duplicating a cube whose edge is one unit leads to the equation $x^3 = 2$, which is an irreducible cubic equation. For if a solution of $x^3 = 2$ were rational, then we could represent the solution by a/b, a and b integers, $b \neq 0$. Let a/b be in lowest terms, i.e., a and b have no common factor greater than 1. Then $a^3 = 2b^3$ and a^3 would be an even integer. Therefore, a would be an even integer, say $2n$, since the cube of an odd integer is odd. Thus $(2n)^3 = 2b^3$

$$b^3 = 4n^3$$

b^3 is even, and b is even, which contradicts the hypothesis that a and b have no common factor greater than 1. Since $x^3 = 2$ is an irreducible cubic equation, its roots cannot be constructed with unmarked straight edge and compasses, and it is not possible to duplicate the cube with those instruments.

Early attempts by Hippocrates and Menaechmus showed that the problem could be solved by finding the intersections of parabolas and hyperbolas. Thus the equations $x^2 = ay$ and $y^2 = bx$ result in the equation $x^3 = 2a^3$, if we let $b = 2a$.

PROBLEM SET IV–A

1. Derive the equation $x^3 = 2a^3$ from $x^2 = ay$ and $y^2 = bx$.
2. Why is this method not considered to be a "solution" of the Delian problem?
3. Show that $x^2 = ay$ and $xy = ab$, with $b = 2a$, also lead to the equation $x^3 = 2a^3$.

Diocles (second century B.C.) used the cissoid to duplicate the cube. Vieta, Descartes, Fermat, and Newton also developed methods for duplicating the cube. Newton used the limaçon of Pascal for this purpose. Of course, none of the methods were restricted to the use of unmarked straight edge and compasses.

PROBLEM SET IV–B

1. Although it is not possible to duplicate a cube, the two-dimensional analogue of the problem can be solved.

Construct a square whose area is twice a given square.

2. Is it possible to construct the radius of a sphere whose surface area is twice the surface area of a unit sphere, or whose volume is twice the volume of a unit sphere?

CHAPTER V

The Problem of Trisecting an Angle

CERTAIN ANGLES CAN be trisected without difficulty. For example, a right angle can be trisected, since an angle of 30° can be constructed. However, there is no procedure, using only an unmarked straight edge and compasses, to construct one-third of an arbitrary angle.

We shall prove this statement by showing that an angle of 60° cannot be trisected. For this purpose we make use of a formula developed in Chapter III,

$$\cos 3\theta = 4 \cos^3 \theta - 3 \cos \theta.$$

Let $3\theta = 60°$, then $\cos 3\theta = \frac{1}{2}$. Also let $x = 2 \cos \theta = 2 \cos 20°$
Then

$$\frac{1}{2} = \frac{x^3}{2} - \frac{3x}{2}$$

33

or

$$x^3 - 3x - 1 = 0$$

We can show that the cubic equation $x^3 - 3x - 1 = 0$ is irreducible. For, suppose $x = a/b$, where a and b are integers with no common factor greater than 1, $b \neq 0$. Then $(a^3/b^3) - (3a/b) - 1 = 0$ and $a^3 - 3ab^2 = b^3$, or $b^3 = a(a^2 - 3b^2)$ and a divides b^3.

Since a and b have no common factor greater than 1, a must be $+1$ or -1. Similarly, $a^3 = b^3 + 3ab^2 = b^2(b + 3ab)$ and b^2 divides a^3. Therefore b is also $+1$ or -1.

Thus the only possible rational roots of $x^3 - 3x - 1 = 0$ are $+1$ or -1. Since neither $+1$ nor -1 is a root of the equation, $x^3 - 3x - 1 = 0$ is an irreducible cubic equation, and its roots cannot be constructed with straight edge and compasses.

Therefore cos 20° cannot be constructed with straight edge and compasses. Since an angle can be constructed if and only if its cosine can be constructed, we have shown that an angle of 20° is not constructible, and that the general angle cannot be trisected.

An alternative method for showing that the cubic $x^3 - 3x - 1 = 0$ is irreducible is to use the following theorem from algebra. If the equation $c_0 x^n + c_1 x^{n-1} + \cdots + c_n = 0$ (with all the coefficients integers) has a rational root a/b, then a is a factor of c_n and b is a factor of c_0. [A proof of this theorem can be found in *Numbers: Rational and Irrational* by Ivan Niven.] Again we reach the conclusion that ± 1 are the only candidates for rational roots.

PROBLEM SET V–A

1. Using a coordinate system
   (a) construct the cos A given acute angle A.
   (b) same as (a) but A is obtuse.
   (c) construct A given a line whose length is cos A.
2. Using the identity $\cos 3\theta = 4 \cos^3 \theta - 3 \cos \theta$, determine whether each of the following angles can be trisected: (a) 90°, (b) 120°, (c) 180°.

An interesting method for trisecting an angle is attributed to Archimedes.

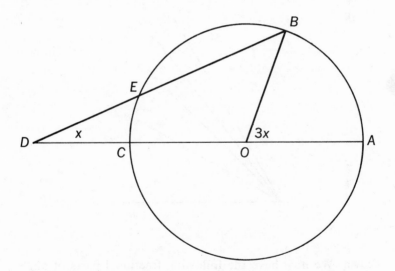

Let AOB be the given angle. Construct a circle with O as center and any convenient length as radius. Construct a line thru B intersecting diameter AC extended so that ED is equal to the radius of the circle. Then angle D is one-third of angle AOB.

PROBLEM SET V–B

1. Prove that angle D is one-third of angle AOB.
2. Why does this method not meet the requirements for constructibility?

Another ingenious procedure for trisecting an angle is as follows:

Let AOB be the given angle. Let OC be the bisector of angle AOB, OD the bisector of angle COB, etc., i.e., continue bisecting the angle formed by OB and the last angle bisector

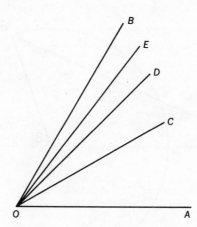

drawn. We now have the following fractional parts of angle AOB—$\frac{1}{2}, \frac{1}{4}, \frac{1}{8}, \frac{1}{16}, \cdots$. Using a pair of compasses and a straight edge, combine the parts of angle AOB, starting with $\frac{1}{4}$

$$\frac{1}{4} + \frac{1}{16} + \frac{1}{64} + \cdots$$

We then have an infinite geometric progression whose ratio is $\frac{1}{4}$ and whose sum is $\frac{1}{3}$.

PROBLEM SET V–C

1. Prove that $\frac{1}{4} + \frac{1}{16} + \frac{1}{64} + \cdots = \frac{1}{3}$
2. Why does this method not meet the requirements for constructibility?

Nicomedes (second century B.C.) used a conchoid for trisecting an angle. Another Greek mathematician, Hippias (fifth century B.C.) invented a curve which he used in trisecting an angle. Since this curve was also used in squaring a circle, it is called a quadratrix. The quadratrix could be used to divide an angle into any number of equal parts. Interested readers may wish to refer to D.E. Smith's *History of Mathematics* for descriptions of the conchoid and quadratrix and how they are used in trisecting an angle.

distance from one end to a line AT. EA perpendicular is elected to
DG meeting the semicircle at G, $EG = x$.

To construct a square equal to a given quadrilateral, one
can proceed as follows: Draw diagonal DB of quadrilateral
$ABCD$. Thru C draw a line parallel to DB and intersecting AB
extended at F. Draw DF. Then triangle AFD is equal in area
to quadrilateral $ABCD$. A square x^2 can be constructed equal in
area to any given triangle by using the equation $x^2 = \frac{1}{2}bh$.

CHAPTER VI

The Problem of Squaring
the Circle

THE GREEKS WERE able to construct a square equal in area to
any given polygon. Thus to construct a square equal to a
given parallelogram, we can use the equation $x^2 = bh$, where
b and h are the base and altitude of the parallelogram, and x
is a side of the square. First construct the altitude of the
given parallelogram. Then a semicircle is constructed with a

PROBLEM SET VI-A

1. Prove that the diagram for the construction of a
square equal to a given parallelogram is $x^2 = bh$.

2. Prove that in the diagram for the construction of a
square equal to a given quadrilateral, triangle AFD is
equal to quadrilateral $ABCD$.

3. Construct a square equal in area to a given quadrilateral. (b) given pentagon.

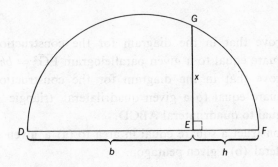

39

diameter equal to $b + h$. At E a perpendicular is erected to DF, meeting the semicircle at G. EG = x.

To construct a square equal to a given quadrilateral, one can proceed as follows: Draw diagonal DB of quadrilateral ABCD. Thru C draw a line parallel to DB and intersecting AB extended at F. Draw DF. Then triangle AFD is equal in area to quadrilateral ABCD. A square can be constructed equal in area to any given triangle by using the equation $x^2 = \frac{1}{2}bh$.

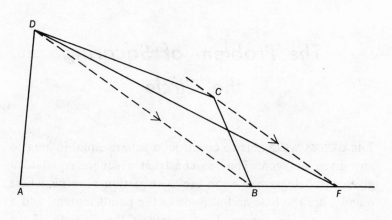

PROBLEM SET VI–A

1. Prove that in the diagram for the construction of a square equal to a given parallelogram $\overline{EG}^2 = bh$.
2. Prove that in the diagram for the construction of a square equal to a given quadrilateral, triangle AFD is equal to quadrilateral ABCD.
3. Construct a square equal in area to (a) a given quadrilateral. (b) a given pentagon.

Since a square can be constructed equal in area to a recti-linear figure, it would be natural to try the same problem for the "simplest" curvilinear figure, a circle. Let us see if we can construct a square equal in area to a given circle of unit radius. The equation becomes $x^2 = \pi$, and we can use the following construction: ADC is a semicircle constructed on AC using $(\pi + 1)$ as a diameter. But how do we determine a length AB $= \pi$?

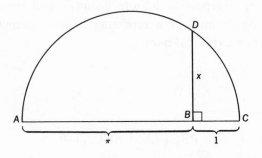

There are three well-defined epochs in the history of attempts to construct π geometrically, or to determine its exact value algebraically. The first period extended from ancient times to the middle of the seventeenth century. It is characterized by ingenious attempts at finding approximate values of π by purely geometric methods. As indicated in a previous chapter, the quadratrix (invented by Hippias) could be used to construct $2/\pi$. The quadratrix could thus be used to square the circle, but the quadratrix itself could not be constructed with straight edge and compasses. By inscribing and circumscribing regular polygons in, and around, a circle, Archimedes (considered to be the greatest of ancient mathematicians) was able to show that $3\frac{1}{7} > \pi > 3\frac{10}{71}$ (a result which has a modern flavor). Archimedes used polygons of 96 sides. Before the end of the first period, Ludolph van Ceulen (sixteenth century) had computed π to 17 places. (In Germany, π is still called the Ludolphian number). In 1621 Snell computed the value of π to 35 places, but he had to use regular polygons of 2^{30} sides.

Still no exact decimal value for π was found, and it was not even known whether such a value could be found—it was not known whether or not π was a rational number. (A rational number can be expressed either as a terminating decimal or a repeating infinite decimal, and conversely, a terminating decimal or a repeating infinite decimal can be expressed as a rational number.)

PROBLEM SET VI–B

1. Determine upper and lower limits for π by inscribing and circumscribing regular polygons of n sides in and about a unit circle, and finding approximations for the circumference of the circle. Use n (number of sides) equal to (a) 4. (b) 8.
2. By carrying out the division to as many places as necessary, show that $3\frac{1}{7}$ is a repeating decimal.
3. Express $3.\overline{14}$ as a rational number. The bar over the 14 means that those two digits repeat indefinitely.

The Greeks were spurred on in their search for a solution to the problem of squaring the circle by a discovery of Hippocrates (fifth century B.C.). Hippocrates showed that it was possible to construct a square equal in area to a curvilinear figure. Starting with an isosceles right triangle, construct an arc AFB of a circle, with C as center and CB as radius. Then with D (midpoint of AB) as center and DB as radius, construct a semicircle AEB. The area of the crescent AEBFA is equal to the area of a square whose side is equal in length to

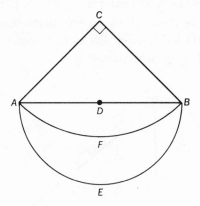

BD. The fact that a square could be constructed equal in area to a crescent led the Greek mathematicians to the belief that the squaring of a circle should not be too difficult. However, the first period ended without any hint as to how the problem could be solved, or whether, in fact, there was a solution.

The second period begins in the second half of the seventeenth century. With the aid of the new analysis, calculus, men like Newton, Fermat, Wallis, and Euler attacked the problem. Numerous expressions for π—involving infinite series, products, and continued fractions—were discovered. For illustrations of these methods of approximating π, the reader can refer to the *History of Mathematics* by D. E. Smith, *The Lore of Large Numbers* by Philip J. Davis, and *Continued Fractions* by C. D. Olds. In 1873 the English mathematician, Shanks, computed π to 707 places, although only 527 were subsequently proved correct.

In 1685 a Jesuit mathematician, Adam Kochansky, gave the following construction for squaring the circle: Construct a circle of unit radius tangent to line RS at A. Let O be the center of the circle. Keep the compasses fixed for the entire construction. With A as center draw an arc cutting the circle in C. With C as center draw an arc intersecting the first arc at D. Draw OD intersecting RS at E. Construct $\overline{EH}$ to be 3 units. Then BH = π (approximate).

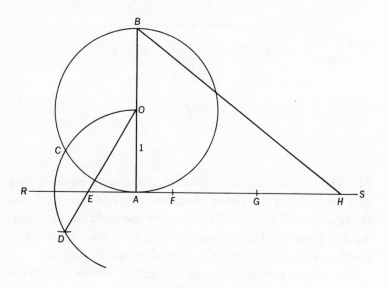

PROBLEM SET VI–C

1. Prove that BH = $\sqrt{\frac{40}{3} - 2\sqrt{3}}$
2. Evaluate the above radical to 5 decimal places and compare your result with the value of π to 5 decimal places.

In 1849 Jakob de Gelder, using an approximation for π obtained by continued fractions, was able to construct a line segment whose length was very close to π. The approximation used by de Gelder was

$$\frac{355}{113} = \left(3 + \frac{4^2}{7^2 + 8^2}\right) = 3.141592\cdots.$$

For details concerning the method of continued fractions, and the geometric construction based upon it, the reader may refer to C. D. Olds' *Continued Fractions*.

These efforts, although they materially increased the accuracy with which π could be expressed, revealed nothing new concerning the fundamental nature of π; e.g., whether or not it was a rational number. But in 1761, the German mathematician, Lambert, proved that π is irrational, that it could not be expressed as a fraction or terminating decimal, or an infinite repeating decimal. Even though Lambert's proof put an end to attempts to find a rational value for π, the question of squaring the circle was as yet unsolved, since there are many irrational numbers (e.g., $\sqrt{2}$) which can be constructed with straight edge and compasses.

However, during this period, Euler made a fundamental discovery which led, about a century and a half later, to a final solution of the problem. Using the newly invented calculus, mathematicians were able to obtain infinite series expansions for various functions.

$$\sin x = x - \frac{x^3}{3!} + \frac{x^5}{5!} - \cdots \qquad (x \text{ in radian measure})$$

$$\cos x = 1 - \frac{x^2}{2!} + \frac{x^4}{4!} - \cdots$$

$$e^x = 1 + x + \frac{x^2}{2!} + \frac{x^3}{3!} + \frac{x^4}{4!} + \cdots$$

Then $\qquad e^{ix} = 1 + ix - \frac{x^2}{2!} - \frac{ix^3}{3!} + \frac{x^4}{4!} + \cdots$

Thus we can show that $e^{ix} = \cos x + i \sin x$. We can then represent the complex number

$$r\,(\cos \theta + i \sin \theta) \text{ as } re^{i\theta}$$

If we let $R_1 = \cos(2\pi/n) + i \sin(2\pi/n)$ (an nth root of unity) then

$$R_1 = e^{i(2\pi/n)}$$
$$R_2 = e^{i(4\pi/n)} = [e^{i(2\pi/n)}]^2 = R_1^2$$
$$R_{n-1} = e^{2i(n-1/n)\pi} = R_1^{n-1}$$
$$R_n = e^{2\pi i} = R_1^n = 1$$

which shows once again that the n nth roots of unity can be written in the form R, R^2, R^3, $\cdots$, $R^n = 1$, where $R = \cos(2\pi/n) + i \sin(2\pi/n)$.

Referring once more to the equation $e^{ix} = \cos x + i \sin x$, let $x = \pi$, then $e^{\pi i} = \cos \pi + i \sin \pi = -1$. The equation $e^{\pi i} = -1$ is one of the most amazing formulas in all of mathematics, relating as it does, the numbers e, π, i, and 1. This relation was used by Lindemann to help in the final solution of the problem of squaring the circle.

PROBLEM SET VI–D

1. Using two terms of the series for $\sin x$ and $\cos x$, compute $\sin \pi/6$ and $\cos \pi/6$ to two decimal places, and compare your results with the values of $\sin \pi/6$ and $\cos \pi/6$ as found in a table. (Use $\pi = 3.14$.)
2. Using five terms of the series for e^x, compute the value of e to two decimal places.

In the third period, the full power of modern analysis was brought to bear on the problem. In 1873, Hermite proved that e is a transcendental number. An algebraic number is a root of a polynomial equation $a_0 x^n + a_1 x^{n-1} + \cdots + a_n = 0$, where all the coefficients a_0, a_1, $\cdots$, a_n are integers. A transcendental number is not algebraic; i.e., a transcendental number is not the root of a polynomial equation with integral (or rational) coefficients. As an example, if a_0, a_1, a_2 are any integers, then $a_0 e^2 + a_1 e + a_2$ cannot be equal to 0.

At the beginning of the nineteenth century when the concept of a transcendental number was introduced, mathematicians had not yet proved that any known number was transcendental.

The first transcendental number was found by the French mathematician, Liouville, in 1851. It is expressed by the infinite series $1/10 + 1/10^{2!} + 1/10^{3!} + \cdots$. Liouville found a whole class of numbers which he proved to be transcendental, but π was not included in that class. Liouville also proved that e cannot be the root of a quadratic equation with rational coefficients, i.e., $a_0 e^2 + a_1 e + a_2$ (where all the a's are integers) cannot be 0.

Meanwhile the German mathematician, Georg Cantor, proved in 1880 that almost all real numbers are transcendental. The final step in the solution of the problem of squaring the circle was taken by Lindemann in 1882, when he generalized the result obtained by Hermite and proved that in an equation of the form

$$a_0 + a_1 e^{\rho_1} + a_2 e^{\rho_2} + \cdots = 0$$

the exponents and the coefficients are not only not integers but that they cannot all be algebraic numbers. The reader can find a proof in *Irrational Numbers* by Ivan Niven. If we apply this result to Euler's equation $1 + e^{\pi i} = 0$, since 1 is algebraic, πi is transcendental. Since the algebraic numbers form a field, the product of any two algebraic numbers is algebraic. (For a proof, see *Irrational Numbers*). Inasmuch as i is algebraic (it is a root of the equation $x^2 + 1 = 0$), π must be transcendental, otherwise πi would be algebraic.

Thus π is not the root of any polynomial equation, and cannot be expressed in terms of rational operations and extractions of real square roots of integers. Therefore, it is not possible to construct a square equal in area to a circle of unit radius. Although it is now known that π is irrational, mathematicians are still interested in the distribution of digits in the decimal expansion of π. Consequently, with the aid of high-speed computers, the value of π has been determined to more than 10,000 decimal places. For further details the reader may refer to *The Lore of Large Numbers* by P. J. Davis.

PROBLEM SET VI–E

In Problem Set VI–C, we found that $\sqrt{\frac{40}{3} - 2\sqrt{3}}$ is an approximate value of π. Show that $\sqrt{\frac{40}{3} - 2\sqrt{3}}$ is the root of a 4th degree equation with integral coefficients.

CHAPTER VII

The Problem of Constructing Regular Polygons

ANOTHER INTRACTABLE PROBLEM concerning the circle which occupied the efforts and attention of the greatest mathematicians from antiquity to the nineteenth century is the problem of dividing a circle into equal arcs. Joining the successive points of division by chords produces a regular polygon (a polygon which is both equilateral and equiangular).

The Greeks were able to construct a regular polygon of 2^m sides, where m is an integer greater than 1. They were also able to construct regular polygons of 3 sides and 5 sides. Since an arc of a circle can be bisected using straight edge and compasses, regular polygons of $3 \cdot 2^m$ and $5 \cdot 2^m$ sides can be constructed, where m is any positive integer. Furthermore, the Greeks had proved that, if a regular polygon of a sides and one of b sides can be constructed and a and b are relatively prime (i.e., a and b have no common factor greater than 1), then a regular polygon of $a \cdot b$ sides can be constructed. The proof uses the euclidean algorithm for finding the greatest common factor of two integers. In case the two integers are relatively prime, the greatest common factor is 1 and the algorithm enables us to solve the following problem.

Given two relatively prime integers a and b, there exist two other integers k and l so that $ka + lb = 1$. As pointed out in the chapter on Complex Numbers, to divide a circle into n equal parts is equivalent to constructing an arc (or central angle) whose measure is $2\pi/n$. Thus if we construct regular polygons of a sides and b sides, we can construct arcs whose measures are $2\pi/a$ and $2\pi/b$. Therefore, we can construct arcs whose measures are $2\pi l/a$ and $2\pi k/b$, where k and l are integers such that $ka + lb = 1$. Finally, we can construct an arc whose measure is $2\pi l/a + 2\pi k/b = 2\pi(ka + lb)/ab = 2\pi/ab$, which proves that a regular polygon of $a \cdot b$ sides can be constructed.

Since 3 and 5 are relatively prime, a regular polygon of 15 sides can be constructed, as well as regular polygons of $15 \cdot 2^m$ sides, where m is a positive integer. We can then summarize the achievement of the Greeks, by stating a general formula which indicates which regular polygons the ancient mathematicians could construct. The Greeks could construct a regular polygon of n sides if n is an integer of the form $2^m \cdot P_1^{r_1} \cdot P_2^{r_2}$, where m is any *non-negative* integer and P_1 and P_2 are the distinct primes 3 and 5, and $r = 0$ or 1.

PROBLEM SET VII-A

1. Construct regular polygons of 2^m sides where m equals
   (a) 2. (b) 3. (c) 4.

2. Construct a regular polygon of $3 \cdot 2^m$ sides where m equals (a) 0. (b) 1. (c) 2.

3. Assuming that regular polygons of 3 sides and 5 sides can be constructed, show how to construct a regular polygon of 15 sides by finding two integers k and l such that $3k + 5l = 1$. How would you then construct a regular polygon of 30 sides?

4. According to what you have learned thus far, is it possible

   (a) to construct a regular polygon of $3 \cdot 3$ or 9 sides?
   (b) to construct a regular polygon of $a + b$ sides if regular polygons of a and b sides can be constructed, and a and b are relatively prime? Try to prove your conjecture.

For more than 2,000 years the problem of dividing a circle into equal parts remained as left by the ancient mathematicians. Despite the fact that such eminent mathematicians as Fermat and Euler worked on the problem, no further progress was made until the end of the eighteenth century, when Gauss solved the problem completely in 1796.

E. T. Bell in his *Development of Mathematics* states "The occasion for Gauss' making mathematics his life work was his spectacular discovery at the age of nineteen concerning the construction of regular polygons by means of straight edge and compasses alone." Prior to his discovery Gauss had been considering a career in Philology. D. E. Smith in his *History of Mathematics* quotes Gauss' own record of his discovery.

"The day was March 29, 1796, and chance had nothing to do with it. Before this, indeed, during the winter of 1796 (my first semester at Göttingen), I had already discovered everything relating to the separation of the roots of the equation, $(x^p - 1)/(x - 1) = 0$ into two groups. After intensive consideration of the relation of all the roots to one another on arithmetical grounds, I succeeded during a holiday at Braunschweig, on the morning of the day alluded to (before I got out of bed,), in viewing the relation in the clearest way, so that I could immediately make application to the 17 sides and to the numerical verifications."

Courant and Robbins in their book *What Is Mathematics?* state "He [Gauss] always looked back on the first of his great feats with particular pride. After his death a bronze statue of him was erected in Göttingen; and no more fitting honor could be devised than to shape the pedestal in the form of a regular 17-gon." Why a regular polygon of 17 sides will become clear later.

Let us now examine Gauss' remarkable achievement. At the beginning of the chapter we indicated that the Greeks could construct a regular polygon of n sides, if n is of the form $2^m P_1^{r_1} P_2^{r_2}$, when $P_1 = 3$ and $P_2 = 5$ and $r = 0$ or 1. In Problem Set VII–A, you were asked to construct a regular polygon of 3 sides, but not of 5. First, then, we shall review a method the Greek mathematicians used to construct a regular polygon of 5 sides.

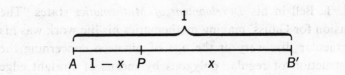

The Greek mathematicians constructed a regular pentagon by dividing a unit line into a mean and extreme ratio. P divides AB into a mean and extreme ratio if the larger segment, x, is the mean proportion between the entire length and the shorter segment; i.e., $1/x = x/(1 - x)$ or $x^2 + x - 1 = 0$. To show how that proportion is related to the regular pentagon, we can proceed as follows.

Let O be a central angle of 36° in a unit circle (central angle subtended by a side of a regular decagon). Then angle A = angle ABO = 72°. Let BD bisect angle ABO.

$$AB = BD = OD = x; \quad AD = 1 - x$$

Since the bisector of the angle of a triangle divides the opposite side into two segments that are proportional to the adjacent sides, $1/x = x/(1 - x)$, and OA has been divided into a mean and extreme ratio. Thus, we have $x^2 + x - 1 = 0$, and $x = (-1 + \sqrt{5})/2$. [Why do we discard $(-1 - \sqrt{5})/2$?]

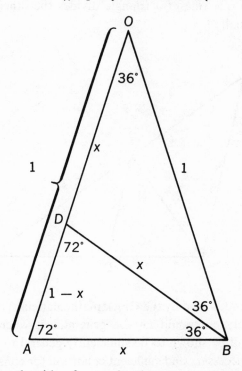

Therefore, the side of a regular decagon can be constructed, and the regular pentagon can be formed by joining the alternate vertices.

PROBLEM SET VII–B

1. Using the above procedure, construct a regular pentagon.
2. Prove that if CD bisects angle ACB, then (AC/CB) = (AD/DB)

[Hint: Thru A draw a line parallel to CD, and let the line intersect BC (extended) at E. Then use the theorem: A line parallel to one side of a triangle divides the other two sides proportionally.]

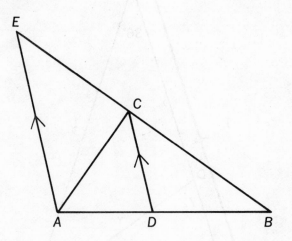

The methods used by the Greek mathematicians to construct regular polygons could not be generalized, whereas Gauss' method will be shown to be sufficiently general to enable one to state a necessary and sufficient condition for constructibility of any regular polygon. First we shall review and extend some of the principles we have used in answering the questions regarding duplicating a cube, trisecting an angle, and squaring a circle.

We have stated that the roots of an irreducible cubic equation (cubic equation with no rational root) cannot be expressed in terms of rational operations and extractions of real square roots performed on the coefficients of the equations. We shall now generalize this theorem. A polynomial equation with rational coefficients, $f(x) = 0$, is said to be irreducible in the field of rational numbers if $f(x)$ cannot be resolved into two rational polynomial factors of degree equal to or greater than 1. It can then be proven that the roots of an irreducible equation of degree n are expressible in terms of rational operations and extractions of real square roots performed on the coefficients of an equation only if n is of degree 2^h, h being a non-negative integer. The reader may refer to *Famous Problems* by F. Klein *et al.* for a proof of that theorem. Therefore, if the coefficients of an irreducible equation represent the lengths of given line segments, and the degree of the equation is not equal to 2^h, then the roots of the equation cannot be constructed with straight edge and compasses.

We have also learned that the n nth roots of unity can be found by solving the equation $x^n - 1 = 0$. If we divide $(x^n - 1)$ by $(x - 1)$, we obtain $x^{n-1} + x^{n-2} + \cdots + x + 1$. The equation $x^{n-1} + x^{n-2} + \cdots + x + 1 = 0$ is called the cyclotomic equation, and it has the same roots as $x^n - 1 = 0$, except for the root $x = 1$. Furthermore, the n nth roots of unity form a group, and can be arranged in the following sequence: R, R^2, R^3, $\cdots$, $R^n = 1$, where $R = \cos(2\pi/n) + i \sin(2\pi/n)$ and R^{-1} or $1/R = \cos(2\pi/n) - i \sin(2\pi/n)$. Finally, the roots of unity, when plotted in the complex plane will divide the unit circle into n equal arcs. Thus, we can conclude that a regular polygon of n sides is constructible only if the cyclotomic equation is an irreducible equation of degree 2^h. Consequently $n - 1 = 2^h$ or $n = 2^h + 1$.

Another theorem we shall need is a generalization of the relation between the roots and coefficients of a quadratic equation. If r_1 and r_2 are the roots of $x^2 + bx + c = 0$, then $r_1 + r_2 = -b$ and $r_1 \cdot r_2 = c$. In general it can be proven that if $r_1, r_2, r_3, \cdots, r_n$ are the roots of the equation $x^n + a_1 x^{n-1} + a_2 x^{n-2} + \cdots + a_{n-1}x + a_n = 0$, then

$$\sum_{i=1}^{n} r_i = -a_1$$

$$\sum_{i,j=1}^{n} r_i r_j = a_2 \qquad (i < j)$$

$$\sum_{i,j,k=1}^{n} r_i r_j r_k = -a_3 \qquad (i < j, i < k, j < k)$$

$$\vdots$$

$$r_1 r_2 \cdots r_n = (-1)^n a_n$$

$\sum r_i r_j$ means the sum of all terms formed by multiplying any two roots together.

PROBLEM SET VII–C

1. If the roots of $x^3 + a_1 x^2 + a_2 x + a_3 = 0$ are r_1, r_2, r_3, then we can write the equation in the form $(x - r_1) \cdot (x - r_2)(x - r_3) = 0$. By multiplying the factors of the left-hand side of the equation, show that

$$r_1 + r_2 + r_3 = -a_1$$
$$r_1 r_2 + r_1 r_3 + r_2 r_3 = a_2$$
$$r_1 r_2 r_3 = -a_3$$

2. Repeat the process for the equation $x^4 + a_1 x^3 + a_2 x^2 + a_3 x + a_4 = 0$.

Let us now use Gauss' method for several values of n. To investigate the possibility of constructing a regular pentagon we use $n = 5$, and let $R = \cos(2\pi/5) + i \sin(2\pi/5)$.

Therefore $R^5 - 1 = 0$, and (since $R \neq 1$)

$$\frac{R^5 - 1}{R - 1} = R^4 + R^3 + R^2 + R + 1 = 0$$

The cyclotomic equation $R^4 + R^3 + \cdots + 1 = 0$ is irreducible and the degree is of the form 2^h. The roots of this equation can be constructed, and we now proceed to accomplish the construction.

We pair the roots of the cyclotomic equation in the following manner:

$$y_1 = R + \frac{1}{R} = R + R^4 = \left(\cos\frac{2\pi}{5} + i \sin\frac{2\pi}{5}\right)$$
$$+ \left(\cos\frac{2\pi}{5} - i \sin\frac{2\pi}{5}\right)$$
$$= 2 \cos\frac{2\pi}{5}$$
$$y_2 = R^2 + \frac{1}{R^2} = R^2 + R^3 = 2 \cos\frac{4\pi}{5}$$

Therefore

$$y_1 + y_2 = R + R^2 + R^3 + R^4 = -1$$

and

$$y_1 \cdot y_2 = R^3 + R + R^4 + R^2 = -1$$

y_1 and y_2 satisfy the equation $y^2 + y - 1 = 0$ (Why?) and $y_1 = (-1 + \sqrt{5})/2$ (which is a result previously obtained). (Note that since $y_1 = 2 \cos(2\pi/5) \geq 0$, $y_1 = (-1 + \sqrt{5})/2$, and since $y_2 = 2 \cos(4\pi/5) < 0$, $y_2 = (-1 - \sqrt{5})/2$. Thus $\cos 2\pi/5 = (-1 + \sqrt{5})/4$, and a regular pentagon can be constructed (since $\cos A$ can be constructed if, and only if, A can be constructed). We can proceed as follows:

In a circle whose radius is 1, draw two perpendicular diameters. With C (midpoint of OA') as a center and CB as a radius, draw an arc cutting OA at D. Then, if S_n represents a side of a regular polygon of n sides, $S_{10} = OD$ and $S_5 = BD$. Notice that we start with an equation of degree 2^2 and obtain an equation of degree 2.

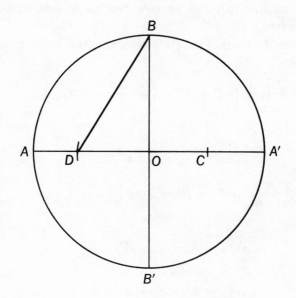

PROBLEM SET VII–D

1. Show that y_1 and y_2 satisfy the equation $y^2 + y - 1 = 0$.
2. Prove that $S_5 = BD$. Hints: First show that $OD = (-1 + \sqrt{5})/2 = S_{10}$. Then prove that in a unit circle $S_5^2 = 1 + S_{10}^2$ or $S_5^2 = S_6^2 + S_{10}^2$.

We shall now investigate the case for $n = 7$. The Greeks and succeeding mathematicians tried to construct a regular heptagon (7-sided polygon) but with no success. It was especially frustrating since regular polygons of 3, 4, 5, 6, 8, 10 sides could be constructed. Why not 7 or 9? To answer that question, let us start with a 7th root of unity, $R = \cos 2\pi/7 + i \sin 2\pi/7$. Then the cyclotomic equation becomes

$$\frac{R^7 - 1}{R - 1} = R^6 + R^5 + R^4 + R^3 + R^2 + R + 1 = 0.$$

The above cyclotomic equation leads to an irreducible cubic equation. Therefore, a regular heptagon cannot be constructed. To show this we pair the roots as follows:

$$y_1 = R + \frac{1}{R} = R + R^6 = 2 \cos \frac{2\pi}{7}$$

$$y_2 = R^2 + \frac{1}{R^2} = R^2 + R^5$$

$$y_3 = R^3 + \frac{1}{R^3} = R^3 + R^4$$

then $y_1 + y_2 + y_3 = R + R^2 + R^3 + R^4 + R^5 + R^6 = -1$ similarly $y_1 y_2 + y_1 y_3 + y_2 y_3 = (R^3 + R + R^6 + R^4) + (R^4 + R^2 + R^5 + R^3) + (R^5 + R + R^6 + R^2) = -2$ and $y_1 \cdot y_2 \cdot y_3 = 1$. Therefore y_1, y_2, y_3 satisfy the equation $y^3 + y^2 - 2y - 1 = 0$.

The only rational roots of this cubic are integers which are divisor's of 1. However, neither $+1$ nor -1 is a root of this equation. Therefore, it is an irreducible cubic and its roots cannot be constructed. Thus, $y_1 = 2[\cos (2\pi/7)]$ cannot be constructed, and it is not possible to construct a regular polygon of 7 sides.

Just as Archimedes described a method for trisecting an angle using a pair of compasses and a straight edge with two marks on it, so he gave a most ingenious method for constructing a regular heptagon using the same instruments. A description of Archimedes' method for constructing the regular heptagon can be found in *Episodes from the Early History of Mathematics* by Asger Aaboe.

For the case $n = 9$, the cyclotomic equation becomes $x^8 + x^7 + \cdots + x + 1 = 0$. Since the degree of the equation is 2^4, it would seem that we should be able to construct its roots and thus to construct a regular polygon of 9 sides.

However, $x^9 - 1$ is not only divisible by $x - 1$, it is also divisible by $x^3 - 1$.

$$\frac{x^9 - 1}{x^3 - 1} = x^6 + x^3 + 1.$$

The equation $x^6 + x^3 + 1 = 0$ has the same roots as $x^9 - 1 = 0$, except for the cube roots of unity. Therefore, if $R = \cos 2\pi/9 + i \sin 2\pi/9$, the roots of the equation $x^6 + x^3 + 1 = 0$ are R, R^2, R^4, R^5, R^7, and R^8. $R^3, R^6, R^9 = 1$ are the roots of $x^3 - 1 = 0$, since $(R^3)^3 = R^9 = 1$, and $(R^6)^3 = R^{18} = 1$. We pair the roots of $x^6 + x^3 + 1 = 0$ as follows:

$$y_1 = R + R^8 = R + \frac{1}{R} = \left(\cos \frac{2\pi}{9} + i \sin \frac{2\pi}{9} \right)$$
$$+ \left(\cos \frac{2\pi}{9} - i \sin \frac{2\pi}{9} \right) = 2 \cos \frac{2\pi}{9}$$

$$y_2 = R^2 + R^7 = R^2 + \frac{1}{R^2} = \left(\cos \frac{4\pi}{9} + i \sin \frac{4\pi}{9} \right)$$
$$+ \left(\cos \frac{4\pi}{9} - i \sin \frac{4\pi}{9} \right) = 2 \cos \frac{4\pi}{9}$$

$$y_3 = R^4 + R^5 = R^4 + \frac{1}{R^4} = \left(\cos \frac{8\pi}{9} + i \sin \frac{8\pi}{9} \right)$$
$$+ \left(\cos \frac{8\pi}{9} - i \sin \frac{8\pi}{9} \right) = 2 \cos \frac{8\pi}{9}$$

Then $y_1 + y_2 + y_3 = R + R^2 + R^4 + R^5 + R^7 + R^8$, which represents the sum of the roots of $x^6 + x^3 + 1 = 0$. Thus, $y_1 + y_2 + y_3 = -a$, where a is the coefficient of x^5; and $y_1 + y_2 + y_3 = 0$. Similarly, $y_1 y_2 + y_1 y_3 + y_2 y = -3$ and $y_1 y_2 y_3 = -1$.

Therefore y_1, y_2, y_3 are roots of the equation $y^3 - 3y + 1 = 0$

Since neither $+1$ nor -1 are roots of $y^3 - 3y + 1 = 0$ it is an irreducible cubic equation, and its roots cannot be constructed. Therefore, a regular polygon of 9 sides cannot be constructed with straight edge and compasses.

PROBLEM SET VII–E

1. Show that R^3 and R^6 are roots of the equation $x^2 + x + 1 = 0$, and determine the value of $R^3 + R^6$.
2. Using the information in the problem above show that
   (a) $y_1y_2 + y_1y_3 + y_2y_3 = -3$
   (b) $y_1y_2y_3 = -1$
3. In example (2) of Problem Set V–A, you were asked to determine whether an angle of 120° can be trisected. The problem leads to the same equation as does the problem of constructing polygon of 9 sides. ($y^3 - 3y + 1 = 0$). Show, by geometric considerations, that the two problems are equivalent.

Let us now return to the question raised previously. Even though the degree of the cyclotomic equation $x^8 + x^7 + \cdots + 1 = 0$ is 2^3, we cannot construct all the roots of this equation. We can now see that this is due to the fact that the equation is reducible. From the analysis of the problem of constructing a regular polygon of 9 sides we see that $x^8 + x^7 + \cdots + 1 = (x^2 + x + 1)(x^6 + x^3 + 1)$. The important question then becomes under what condition is the cyclotomic equation irreducible, under what condition is it reducible. If we examine the two cases we have considered thus far, where n is of the form $2^h + 1$ ($n = 5$, $n = 9$), we notice that when n is prime, the construction is possible, when n is composite the construction is not possible. In fact it is possible to prove that when n is of the form $2^h + 1$ and n is prime, then the cyclotomic equation is of degree 2^h and is irreducible and the unit circle can be divided into n equal parts; but, on the contrary, if n is of the form $2^h + 1$, and n is not prime, the cyclotomic equation is reducible and the division of the unit circle is not possible with straight edge and compasses.

We can go one step further by proving that if $2^h + 1$, (h being a non-negative integer) is a prime number, then $h = 2^m$ (where m is a non-negative integer). Since, if h has an odd factor greater than 1, h can be written in the form rs, r and s being positive integers, s an odd integer > 1. Then $2^h + 1 = (2^r)^s + 1$. Now $a^s + b^s$ can be factored if s is odd. $a^s + b^s = (a + b)(a^{s-1} - a^{s-2}b + a^{s-3}b^2 - \cdots + b^{s-1})$. Thus $a^3 + b^3 = (a + b)(a^2 - ab + b^2)$ and $(2^r)^s + 1 = (2^r + 1)$ $\cdot (2^{r(s-1)} - 2^{r(s-2)} + 2^{r(s-3)} - \cdots + 1)$. Since r and s are positive integers and $s \geq 2$, $2^h + 1$ will have two factors each > 1, which is contrary to the hypothesis that $2^h + 1$ is a prime; consequently if $2^h + 1$ is a prime it is of the form $2^{2^m} + 1$, m a non-negative integer.

The number $2^{2^m} + 1$ has turned up on many occasions in the history of mathematics. It is known as the "Fermat" number, after Pierre Fermat (1601–1665), a French mathematician of the first rank. Euclid gave an ingenious proof that there are an infinite number of primes. Ever since mathematicians have been striving to find a formula that would always give a prime number, even though not all prime numbers.

Fermat conjectured that all numbers of the form $2^{2^m} + 1$ were prime. Even though he was convinced of the truth of this statement, he indicated that he could not prove it. In fact for

$m = 0, 1, 2, 3, 4, 2^{2^m} + 1$ is respectively 3; 5; 17; 257; 65, 537 (each of these numbers is prime). But Euler showed that $2^{2^5} + 1$ can be factored. It can be shown that 641 is a divisor of $2^{32} + 1$, by the following procedure: $641 = 5 \cdot 2^7 + 1$. $\therefore$ 641 divides $(5 \cdot 2^7 + 1)(5 \cdot 2^7 - 1) = 5^2 \cdot 2^{14} - 1$ and 641 divides $(5^2 \cdot 2^{14} - 1)(5^2 \cdot 2^{14} + 1) = 5^4 \cdot 2^{28} - 1$. But 641 is also equal to $5^4 + 2^4$. $\therefore$ 641 also divides $5^4 \cdot 2^{28} + 2^{32}$. Consequently 641 divides $(5^4 \cdot 2^{28} + 2^{32}) - (5^4 \cdot 2^{28} - 1) = 2^{32} + 1$.

PROBLEM SET VII–F

1. By actual division show that 641 is a factor of $2^{2^5} + 1$.

To summarize, we can now state that a regular polygon of n sides can be constructed if n is a prime of the form $2^{2^m} + 1$. We have already shown how to construct the regular polygons for $m = 0$ and $m = 1$. For $m = 2$, $2^{2^2} + 1 = 17$. In the 2,000 years between Euclid and Gauss it was not even suspected that a regular polygon of 17 sides could be constructed. Now you can see why Gauss desired to have the base of the "pedestal in the form of a regular 17-gon."

We shall now outline the procedure Gauss used in constructing the regular 17-gon. The cyclotomic equation is $R^{16} + R^{15} + \cdots + R + 1 = 0$. In order to pair the roots of this equation we first find an integer g, such that all the roots can be arranged in the order $R, R^g, R^{g^2}, \cdots$ where R is a *primitive* 17th root of unity. R is a primitive nth root of unity if $R^n = 1$ and $R^e \neq 1$ for all positive integers $e < n$.

PROBLEM SET VII–G

1. Show that $g = 2$ will not give all the roots of $R^{16} + R^{15} + \cdots + R = 0$; $R, R^2, \cdots, R^{16}$.

2. Show that $g = 3$ will give all the roots in the following order:

$$R, R^3, R^9, R^{10}, R^{13}, R^5, R^{15}, R^{11}, R^{16}, R^{14}, R^8, R^7, R^4,$$
$$R^{12}, R^2, R^6.$$

We select alternate terms in the sequence in Problem Set VII–G (2), and obtain

$$y_1 = R + R^9 + R^{13} + R^{15} + R^{16} + R^8 + R^4 + R^2$$
$$y_2 = R^3 + R^{10} + R^5 + R^{11} + R^{14} + R^7 + R^{12} + R^6$$
$$y_1 + y_2 = -1$$

and

$$y_1 y_2 = -4$$

y_1 and y_2 satisfy the equation $y^2 + y - 4 = 0$. We take alternate terms in y_1

$$z_1 = R + R^{13} + R^{16} + R^4, \quad z_2 = R^9 + R^{15} + R^8 + R^2$$

and alternate terms in y_2

$$w_1 = R^3 + R^5 + R^{14} + R^{12}, \quad w_2 = R^{10} + R^{11} + R^7 + R^6$$

then,

$$z_1 + z_2 = y_1 \qquad w_1 + w_2 = y_2$$
$$z_1 \cdot z_2 = -1 \qquad w_1 \cdot w_2 = -1$$

and,

$$z^2 - y_1 z - 1 = 0 \quad w^2 - y_2 w - 1 = 0$$

Finally we take alternate terms in z_1

$$v_1 = R + R^{16}, \qquad v_2 = R^{13} + R^4$$

then,

$$v_1 + v_2 = z_1$$
$$v_1 \cdot v_2 = w_1$$

and,

$$v_1, v_2 \text{ satisfy } v^2 - z_1 v + w_1 = 0$$
$$R, R^{16} \text{ satisfy } r^2 - v_1 r + 1 = 0$$

Thus we can find R by solving a series of quadratic equations.

But there are 16 possible values of R, since there are 16 primitive 17th roots of unity. We would like R to be $(\cos 2\pi/17) + i \sin (2\pi/17)$. Then $1/R$ becomes $(\cos 2\pi/17) - i \sin (2\pi/17)$ and $v_1 = R + 1/R = 2 \cos (2\pi/17)$, $v_2 = R^4 + 1/R^4 = 2 \cos (8\pi/17)$. Since $2\pi/17$ and $8\pi/17$ are both less than $\pi/2$; and, in the first quadrant, the cosine of an angle decreases as the number of degrees in the angle increases, we can see that

$$v_1 > v_2 > 0. \quad \text{Thus} \quad z_1 = v_1 + v_2 > 0.$$

Similarly

$$w_1 = \left(R^3 + \frac{1}{R^3} \right) + \left(R^5 + \frac{1}{R^5} \right) = 2 \cos \frac{6\pi}{17} + 2 \cos \frac{10\pi}{17}$$

$$= 2 \cos \frac{6\pi}{17} - 2 \cos \frac{7\pi}{17}$$

Since

$$\left| \cos \frac{6\pi}{17} \right| > \left| \cos \frac{7\pi}{17} \right|, \, w_1 > 0$$

also

$$y_2 = \left(R^3 + \frac{1}{R^3} \right) + \left(R^5 + \frac{1}{R^5} \right) + \left(R^6 + \frac{1}{R^6} \right) + \left(R^7 + \frac{1}{R^7} \right)$$

$$= 2 \cos \frac{6\pi}{17} + 2 \cos \frac{10\pi}{17} + 2 \cos \frac{12\pi}{17} + \frac{14\pi}{17}$$

The only positive term in y_2 is the first term and

$$\left| \cos \frac{6\pi}{17} \right| < \left| \cos \frac{5\pi}{17} \right| = \left| \cos \frac{12\pi}{17} \right|$$

Therefore $y_2 < 0$.

Since $y_1y_2 = -4$, $y_1 > 0$.

As a result we have now determined the following values:

$$y_1 = \tfrac{1}{2}(\sqrt{17} - 1)$$
$$y_2 = \tfrac{1}{2}(-\sqrt{17} - 1)$$
$$z_1 = \tfrac{1}{2}y_1 + \sqrt{1 + \tfrac{1}{4}y_1^2}$$
$$w_1 = \tfrac{1}{2}y_2 + \sqrt{1 + \tfrac{1}{4}y_2^2}$$

By using the Pythagorean theorem these four lengths can be constructed, and the roots of the equation $v^2 - z_1v + w_1 = 0$ can be constructed. Since the larger root $v_1 = 2\cos(2\pi/17)$, it is possible to construct a regular polygon of 17 sides.

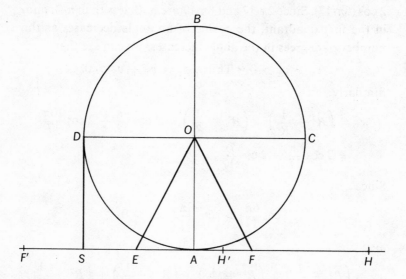

One method of constructing the regular polygon of 17 sides is as follows: In a circle of unit radius, construct two perpendicular diameters AB and CD. Let the tangents at A and D intersect at S. Divide AS into four equal parts, and let AE $= \tfrac{1}{4}$AS. Let the circle with center at E and radius OE cut AS at F and F′. Let the circle with center at F and radius FO cut AS at H (outside of F′F), and the circle with center at F′ and radius F′O cut AS at H′ (between F′ and F). We can then prove that AH $= z_1$ and AH′ $= w_1$.

Finally, by using the method for constructing the roots of a quadratic equation (explained in Chapter I), we can construct v_1, the larger root of $v^2 - z_1v + w_1 = 0$. Construct a circle whose diameter BD joins the points B $(0, 1)$ and D (z_1, w_1). Then the abscissas of G and F (the points where the circle intersects the X-axis) will be the roots of $v^2 - z_1v + w_1 = 0$. $v_1 = 2 \cos (2\pi/17) = $ OF.

[The same unit must be used for all coordinate systems employed in this construction.]

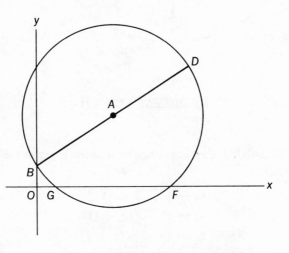

Finally we construct a side of a regular 17 sided polygon: On the X-axis mark off OF $= v_1$. Let M be the midpoint of OF. Construct MP⊥OF. Then AP is one side of a regular polygon of 17 sides.

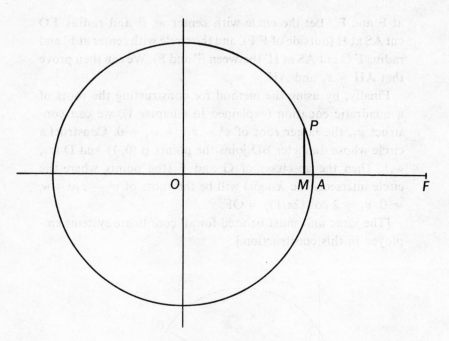

PROBLEM SET VII–H

1. By solving the appropriate quadratic equations, show that

$$y_1 = \tfrac{1}{2}(\sqrt{17} - 1)$$
$$y_2 = \tfrac{1}{2}(-\sqrt{17} - 1)$$
$$z_1 = \tfrac{1}{2}y_1 + \sqrt{1 + \tfrac{1}{4}y_1^2}$$
$$w_1 = \tfrac{1}{2}y_2 + \sqrt{1 + \tfrac{1}{4}y_2^2}$$

2. Prove, by using Fig. VII–1,

   (a) $OE = \tfrac{1}{4}\sqrt{17}$

   (b) $AF = \tfrac{1}{4}\sqrt{17} - \tfrac{1}{4} = \tfrac{1}{2}y_1$

   (c) $AF' = \tfrac{1}{4}\sqrt{17} + \tfrac{1}{4} = -\tfrac{1}{2}y_2$

   (d) $OF = \sqrt{1 + \tfrac{1}{4}y_1^2}$, $OF' = \sqrt{1 + \tfrac{1}{4}y_2^2}$

(e) $AH = \frac{1}{2}y_1 + \sqrt{1 + \frac{1}{4}y_1^2} = z_1$

(f) $AH' = \frac{1}{2}y_2 + \sqrt{1 + \frac{1}{4}y_2^2} = w_1$

3. Prove, using Fig. VII–2 that $OF = v_1 = 2\cos{(2\pi/17)}$.
4. Using Fig. VII–3, show that angle $MOP = 2\pi/17$.
5. Starting with a unit circle, construct
   (a) z_1 and w_1
   (b) v_1
   (c) an angle whose measure is $2\pi/17$
   (d) a regular polygon of 17 sides

The procedure can be generalized as follows:

If n is a prime of the form $2^{2^m} + 1$ or $2^h + 1$, the $(n-1)$ imaginary nth roots of unity can be separated into two sets each of 2^{h-1} roots, each of these sets subdivided into two sets each of 2^{h-2} roots, etc., until we reach the pairs R, $1/R$; R^2, $1/R^2$, etc. We shall then have a series of quadratic equations, the coefficients of any one of which depend only on the roots of the preceding equation in the series. Thus, the roots of $x^n = 1$ can be found in terms of a finite number of rational operations and extractions of real square roots, and a regular polygon of n sides can be constructed if n is a prime of the form $2^{2^m} + 1$.

We can now state a formula which will tell us exactly which regular polygons are constructible. An n-gon is constructible if and only if n is of the form $2^s \cdot p_1^{r_1} \cdot p_2^{r_2} \cdots$, where s is a nonnegative integer, p_1, p_2, $\cdots$ are distinct primes of the form $2^{2^m} + 1$, and each $r = 0$ or 1.

PROBLEM SET VII–K

List the 24 regular polygons with number of sides $n \leq 100$, which can be constructed.

There is still one unresolved question—for which values of m is $2^{2^m} + 1$ a prime number? We know that Fermat's number is prime for $m = 0, 1, 2, 3, 4$, and in fact, we have constructed regular polygons of 3, 5, and 17 sides. For $m = 3$, $n = 257$; and for $m = 4$, $n = 65,537$, the analysis has also been accomplished. L. E. Dickson, in a discussion of "Constructions With Ruler and Compasses" which appears in *Monographs on Topics of Modern Mathematics*, states "The regular 257-gon has been discussed at great length by Richelot in Crelle's *Journal für Mathematik*, 1832; and geometrically by Affolter and Pascal in *Rindicinti della R. Accademia di Napoli*, 1887.

The regular polygon of $2^{16} + 1 = 65,537$ sides has been discussed by Hermes; *Gottingen Nachridten*, 1894."

Also, in the April 1961 issue of *Scientific American*, Martin Gardner describes some of the topics discussed by H. M. S. Coxeter in a recently published book *An Introduction to Geometry*. Martin Gardner quotes Professor Coxeter to the effect that there is at the University of Göttingen a large box containing a manuscript showing how to construct a regular polygon of 65,537 sides. Gardner also writes "a polygon with a prime number of sides can be constructed in the classical manner only if the number is a special type of prime called a Fermat prime; a prime that can be expressed as $2^{2^n} + 1$. Only five such primes are known—3, 5, 17, 257, 65,537. The poor fellow who succeeded in constructing the 65,537-gon, Coxeter tells us, spent ten years in the task."

As Euler showed, when $m = 5$, $2^{2^m} + 1$ has a factor 641. Dickson states that for $n = 6, 7, 8, 9$, the number is not prime. For the next case $n = 10$, it has not been determined whether or not this Fermat number is prime. It may very well be that $2^{2^m} + 1$ is prime only for values of $m < 5$. In that case the formula indicating which polygons are constructible would read

$$n = 2^s p_1^{r_1} \cdot p_2^{r_2} \cdot p_3^{r_3} \cdot p_4^{r_4} \cdot p_5^{r_5}; \quad r = 0 \text{ or } 1 \text{ and } p_1 = 3$$
$$p_2 = 5, p_3 = 17, p_4 = 257, \text{ and } p_5 = 65{,}537$$

However, no proof has been obtained which shows that $2^{2^m} + 1$ is composite for $m \geq 5$; and until this question is answered, we cannot state that the problem of constructing regular polygons has been completely solved.

Dickson concludes his monograph on "Constructions with Ruler and Compasses" by stating "The proof that a regular polygon of p sides, where p is a prime of the form $2^h + 1$ is geometrically inscriptible in a circle was first made by Gauss, *Desquisitiones Arithmeticae*, translated in German by Mach [note: Gauss, although German, wrote in Latin, as did his illustrious predecessors, Newton and Euler.] On page 447 of the latter, Gauss states that a regular n-gon is not inscriptible if n contains an odd prime factor *not* of the form $2^h + 1$, or the square of a prime $2^h + 1$, but no proof appears to have been published by Gauss." Thus it appears Gauss established that a sufficient condition for a regular polygon of n sides to be constructed is that $n = 2^s \cdot p_1^r \cdot p_2^r \cdot \cdots$, where $p_1, p_2 \cdots$ are distinct primes of the form $2^{2^m} + 1$. However, there is no evidence that he proved this condition necessary for the regular polygon to be constructible.

PROBLEM SET VII–L

1. How many digits does the number $2^{2^{10}} + 1$ have? [Hint: Use the fact that $\log_{10} 2 = .301$.]
2. If n is the number of degrees in an angle which can be constructed with straight edge and compasses, and n is an integer, show that $n \geq 3$.

CHAPTER VIII

Concluding Remarks

THE HISTORY OF CONSTRUCTIONS with straight edge and compasses shows quite clearly that all branches of elementary mathematics (algebra, geometry, trigonometry, and analysis) are closely related. The number system of mathematics runs like a unifying thread throughout the entire story. A diagram in *Numbers; Rational and Irrational* by Niven summarizes the situation very aptly (see next page).

The constructible numbers being the roots of irreducible equations of degree 2^h, h, a non-negative integer, consist of all numbers expressible in terms of rational operations and the extractions of real square roots. Yet the imaginary numbers played a crucial role in solving the problem of constructibility of regular polygons.

Any real number can be expressed as an infinite decimal. If the constructible number is rational it can be expressed as a repeating decimal. It is interesting to note that if the con-

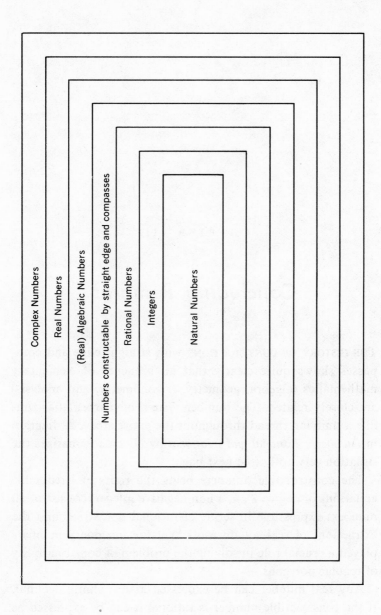

Complex Numbers

Real Numbers

(Real) Algebraic Numbers

Numbers constructable by straight edge and compasses

Rational Numbers

Integers

Natural Numbers

74

structible number is a quadratic irrational (a number of the form $(p \pm \sqrt{D})/Q$, P, Q, D integers, $Q \neq 0$, $D > 0$ and not a perfect square), then the number also has a periodic expansion, not as a decimal, of course, but as a continued fraction. In 1770 Lagrange proved the following theorem: "Any quadratic irrational number has a continued fraction expansion which is periodic from some point onward." (See *Continued Fractions* by Olds). Conversely, any infinite repeating decimal can be expressed as a rational number, and any infinite repeating continued fraction can be expressed as a quadratic irrational. Thus there is a close connection between constructibility and periodicity.

Many people have the mistaken notion that when a mathematician states that a certain construction (e.g., trisecting an angle of 60°) is impossible, that no solution to the problem has yet been found. "Angle trisectors" and "circle squarers" still exist. But we know their efforts are futile, since proving the impossibility of a certain construction is just as much a solution as proving that a construction is possible and then exhibiting that construction.

There have been other occasions in the history of mathematics when it turned out that a certain result being sought was found to be impossible. Gauss proved that every polynomial equation has a complex root. Ancient mathematicians were able to solve all linear and quadratic equations, As a matter of fact the formulae giving the roots of these equations were expressed in terms of rational operations and extractions of roots performed upon the coefficients. Thus the formula for solving the linear equation $ax + b = c(a \neq 0)$ is $x = (c - b)/a$, and the formula for solving the quadratic equation $ax^2 + bx + c = 0(a \neq 0)$ is

$$x = \frac{-b \pm \sqrt{b^2 - 4ac}}{2a}$$

That was how matters stood until the sixteenth century when

a number of Italian mathematicians (Cardan, Tartaglia, and Ferrari) obtained formulas for the solutions of the cubic (3rd degree) and quartic (4th degree) equations. Again the formulas involved only rational operations and the extractions of roots (cube root for the cubic equation; and fourth root for the quartic equation). Consequently, we say that equations of degree one to four inclusive can be "solved by radicals."

Mathematicians were thus encouraged to seek a formula which would give the roots of any quintic (5th degree) equation. It was only natural to expect that the formula should involve rational operations and the extractions of a 5th root. But all attempts to find such a solution failed. Finally, the Norwegian mathematician, Abel (1802–1829), proved that an equation of degree five cannot be solved in terms of radicals. Shortly thereafter, the French mathematician, Galois, proved the general theorem that an algebraic equation of degree $n \geq 5$ cannot be "solved" by radicals. In the process of proving this theorem, Galois developed the theory of groups of substitutions, which forms an important part of modern algebra.

Even the Greek mathematicians had already established the impossibility of certain "constructions" involving the number system. Euclid proved that it is impossible to have a largest prime number; i.e., given any prime number p, there exists a larger prime number. The Greeks also proved that the ratio of the diagonal of a square to a side $(\sqrt{2})$ cannot be written in the form a/b, a and b integers. It took a couple of thousand years to prove the same theorem for the ratio of the circumference of a circle to its diameter (π).

To illustrate the difference between an "unsolved" problem and a problem which has a solution of the type discussed, we shall consider several examples of unsolved problems. We have already encountered an unsolved problem, namely: are there any prime numbers of the form $2^{2^h} + 1$ where $h > 4$?

Another unsolved problem is *Goldbach's conjecture*—every even number greater than two is the sum of two primes (1 is not considered a prime number; therefore the restriction "greater than two"). Goldbach had proposed this problem in a letter to Euler in 1742. Euler was unable to give a solution, and even to this day there is no complete solution for this problem.

One of the most famous of unsolved problems, which can be stated in terms of elementary concepts, is known as *Fermat's Last Theorem*. The problem states "Are there any positive integers x, y, z, such that $x^n + y^n = z^n$, where n is an integer greater than 2. Of course if $n = 2$ then $x^2 + y^2 = z^2$ has many integral solutions, e.g., $x = 3$, $y = 4$, $z = 5$ or $x = 5$, $y = 12$, $z = 13$. However, for $n = 3$, it is impossible to find integral solutions of $x^n + y^n = z^n$. *Hungarian Problem Book II*, (p. 31), in discussing the solution to a problem, makes the following comments on Fermat's Last Theorem: "The statement is known to be true for $n < 2003$, and as of 1961 for all prime exponents less than 4002. In spite of the efforts of many distinguished mathematicians, no proof has been found for Fermat's conjecture to this day."

Fermat maintained that he had a most wonderful proof for this conjecture, but that the margin of the book he was reading (a book on number theory by the Greek mathematician, Diophantus) was too small to contain the proof. Fermat's "proof" has never been found, and as yet, mathematicians have been unable to supply any, despite the use of more modern and powerful techniques than those at the disposal of Fermat. It is possible that Fermat had a proof which mathematicians would accept as valid, or perhaps there may have been a flaw in Fermat's proof. There have been instances where eminent mathematicians gave proofs that were later found to be defective. The question as to whether Fermat had a proof for his *Last Theorem* is still an unsolved problem.

The difference between an unsolved problem and a problem which has a solution that indicates a certain "construction" to be impossible should thus be clear. There is still one other factor to consider. In 1931 Kurt Gödel proved that in any mathematical system, there may be statements whose truth cannot be decided. Perhaps, Fermat's Last Theorem is one such statement.

There are several possibilities one must consider, in attempting a solution to a problem. One can prove a solution exists, and actually exhibit the solution; one may prove that no solution is possible, or one may prove that the statement is undecidable in terms of the axioms of the mathematical system which provides the frame of reference.

Recently (December 1963) an article appeared in *The New York Times* which illustrated the last possibility. The number of natural numbers is infinite. The number of rational numbers is also infinite. Since the set of rational numbers can be placed into a 1-1 correspondence with the set of natural numbers, we say that the two sets have the same cardinal number. Any infinite set of numbers that can be placed into a 1-1 correspondence with the set of natural numbers is said to be countable or denumerable. Let us represent the cardinal number of such a set by a. Cantor showed that the cardinal number of the set of real numbers (rational and irrational) is greater than a. Let us call this number β. One unsolved problem in set theory has been "Is there any transfinite number greater than a and less than β?" The article in *The New York Times* describes the investigations of a mathematician from Stanford Univ., Paul J. Cohen, who proved that the question of the existence of a number greater than a and less than β is undecidable on the basis of the axioms of set theory. Professor Cohen's achievement has also been described in the January 1964 issue of *Scientific American* (Science and the Citizen).

An Italian mathematician, Mascheroni, in the eighteenth century was able to demonstrate the surprising fact that any construction possible with unmarked straight edge and compasses is also possible by compasses alone. Of course one cannot construct a straight line with compasses alone, but one can say that a straight line is determined when two points that lie on that line have been constructed. Thus a circle has been divided into 17 equal parts using only a single pair of compasses. Such constructions are known as Mascheroni constructions.

Of course, not all possible constructions can be performed with straight edge alone, since the use of the straight edge only enables one to perform those constructions based upon rational operations. The extraction of square roots requires the intersection of two circles or a circle and a straight line, inasmuch as the extraction of a square root implies a quadratic equation; and we have already seen that constructing the roots of a quadratic equation involves the use of a circle. It is therefore most surprising that all constructions that are possible with straight edge and compasses can be performed by straight edge alone, provided we are permitted to use a single circle with a fixed center and radius.

To appreciate how unexpected was this result proved by Jacob Steiner in the nineteenth century, recall the procedure for finding $\frac{1}{2} b$ where b is a given line segment. Even this simple construction when performed in the usual manner involves the use of two separate circles. Constructions using straight edge only require the use of theorems from projective geometry. The interested reader can find a discussion of "ruler constructions" in *Squaring the Circle and other Monographs*—"Ruler and Compasses" by Hilda P. Hudson.

PROBLEM SET VIII-A

Students very often show their geometry teachers the following method for "trisecting" an angle: Mark off equal lengths AB and AC on the sides of the given angle A. Divide line segment BC into three equal parts. Then draw AD and AE, which trisects angle A.

Prove that AD and AE do not trisect angle A.

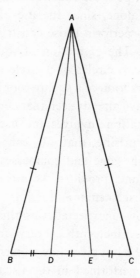

I hope this brief discussion has shown the reader why the solutions to the problems of "squaring the circle and trisecting an angle" are considered to be "profound insights" and "great contributions to human knowledge"; why centuries of "great intellectual effort" were required to solve such seemingly simple problems, and what new mathematics had to be developed to resolve these problems.

SUGGESTIONS FOR FURTHER READING

Men of Mathematics—E. T. Bell
Development of Mathematics—E.T. Bell
What is Mathematics?—Courant and Robbins
New First Course in Theory of Equations—L. E. Dickson
History of Mathematics (2 Vol.)—D.E. Smith
New Mathematical Library
    Volume I—Numbers, Rational and Irrational,—I. Niven
    Volume VI—Lore of Large Numbers,—P. J. Davis
    Volume VII—Uses of Infinity,—L. Zippen
    Volume IX—Continued Fractions,—C. D. Olds
    Volume X—Graphs and Their Uses,—O. Ore
    Volume XII—Hungarian Problem Book II
    Volume XIII—Episodes from the Early History of Mathe-
        matics,—A. Aaboe
Scientific American
    April 1961—Mathematical Games, — M. Gardner
    January 1964—Science and the Citizen—Answers in
        Set Theory
Mathematics Teacher—December 1960
    The Evolution of Extended Decimal Approximations to
    π—J. W. Wrench

MORE ADVANCED BOOKS

A Survey of Modern Algebra,—Birkhoff and MacLane
Modern Algebraic Theories,—L.E. Dickson
The Carus Mathematical Monographs,
    Number II—Irrational Numbers,—I. Niven
Famous Problems and Other Monographs
    1) Famous Problems of Elementary Geometry,—F. Klein
    2) Three Lectures on Fermat's Last Theorem,—L. J. Mordell
Squaring the Circle and Other Monographs
    1) Squaring the Circle,—E. W. Hobson
    2) Ruler and Compass,—H. P. Hudson

Solutions to the Problems

Solutions to the Problems

SET I–A

(1)

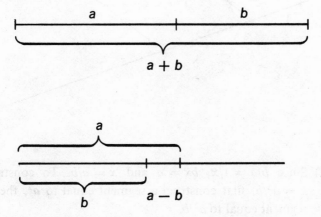

PROBLEM SET I–B

(1) Since a line parallel to one side of a triangle divides the other two sides into proportional segments, we have $1/a = b/x$. Therefore $x = ab$.

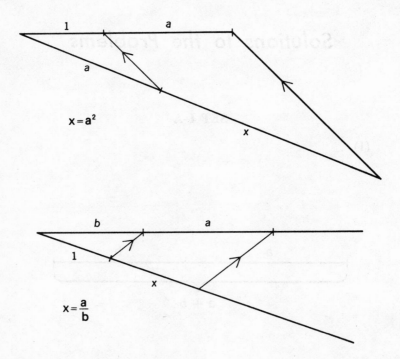

(2) Since $b/a = 1/x$, $bx = a$ and $x = a/b$. To construct $x = a^2/b$, first construct a segment equal to a^2, then a segment equal to a^2/b.

SET I–C

(1) Draw LP and MP. Then triangle LPM is a right triangle and PM is the altitude on the hypotenuse. Therefore $a/x = x/1$ and $x^2 = a$ or $x = \sqrt{a}$.

(2) Instead of using the segments a and 1 for the diameter use the segments a and b. Then $x = \sqrt{ab}$.

To construct $\sqrt[4]{a}$ use $\sqrt{a}$ and 1 as segments for the diameter. Use $\sqrt[4]{a}$ and 1 to construct $\sqrt[8]{a}$.

(3) Construct a right triangle whose legs are 1 and 1 to construct $\sqrt{2}$. For $\sqrt{3}$ use $\sqrt{2}$ and 1 as the lengths of the legs. Use 2 and 1 for $\sqrt{5}$. Use 4 and 1 for $\sqrt{17}$.

SET I–D

$x = (c - b)/a$. First construct $c - b$ then $(c - b)/a$.

SET I–E

(1) The discriminant of the quadratic equation is $a^2 - 4b$. In order for the roots be real and distinct $a^2 - 4b > 0$ or $a^2 > 4b$.

(2) The equation of a circle is $(x - b)^2 + (y - k)^2 = r^2$ where the center is (b, k) and the radius $= r$. The coordinates of A are $(a/2, (b + 1)/2))$, $AB^2 = r^2 = a^2/4 + [(b + 1)/2 - 1]^2 = a^2/4 + (b - 1)^2/4$.   Therefore

$(x - a/2)^2 + (y - (b + 1)/2)^2 = a^2/4 + (b - 1)^2/4$. To find the abscissas of G and F, let $y = 0$ in the equation of the circle and solve for x. The results are $x = (a + \sqrt{a^2 - 4b})/2$ and $x = (a - \sqrt{a^2 - 4b})/2$, which are the roots of $x^2 - ax + b = 0$.

SET I–F

(1) Construct a circle with diameter BD where the coordinates of B are $(0, 1)$ and the coordinates of D are $(-1, -1)$.

SET II–A

a) Solve the equations by elimination of one variable or by determinants.

$$x = \frac{bc' - b'c}{ab' - a'b}$$

$$y = \frac{a'c - ac'}{ab' - a'b}$$

b) Since $y = -(ax + c)/b$.

$$x^2 + \left(\frac{-ax - c}{b}\right)^2 + dx + e\left(\frac{-ax - c}{b}\right) + f = 0.$$

By using the quadratic formula for the equation, $AX^2 + BX + C = 0$ where $A = a^2 + b^2$

$$B = 2ac + b^2d - abe$$
$$C = c^2 - 2bce + b^2f.$$

We find $x = -(B \pm \sqrt{B^2 - 4AC})/2A$, A, B, C being given above. A similar procedure enables us to find y.

c) By subtracting the two equations we find the equation of the line passing through the two points (real or imaginary) of intersection of the two circles $(d - d')x + (e - e')y + (f - f') = 0$. We can then use the method of (b) to find the coordinates of the points of intersection of the two circles.

PROBLEM SET II–B

1) $\dfrac{7}{5 - \sqrt{2}} \cdot \dfrac{5 + \sqrt{2}}{5 + \sqrt{2}} = \dfrac{35 + 7\sqrt{2}}{23} = \dfrac{35}{23} + \dfrac{7}{23}\sqrt{2}$

$$a = \frac{35}{23}, \; b = \frac{7}{23}, \; k = 2$$

2) $\dfrac{5}{2 - \sqrt[4]{3}} \cdot \dfrac{2 + \sqrt[4]{3}}{2 + \sqrt[4]{3}} = \dfrac{10 + 5\sqrt[4]{3}}{4 - \sqrt[4]{9}} \cdot \dfrac{4 + \sqrt[4]{9}}{4 + \sqrt[4]{9}}$

$$= \frac{40 + 20\sqrt[4]{3} + 10\sqrt[4]{9} + 5\sqrt[4]{27}}{13}$$

$$= \left(\frac{40 + 10\sqrt{3}}{13}\right) + \left(\frac{20 + 5\sqrt{3}}{13}\right)\sqrt[4]{3}$$

$$a_1 = \frac{40 + 10\sqrt{3}}{13}, \; b_1 = \frac{20 + 5\sqrt{3}}{13}, \; k_1 = \sqrt{3}$$

PROBLEM SET II–C

1) Let $x = \sqrt{3} + \sqrt[4]{2}$
   $x^2 - 3 = \sqrt[4]{2}$
   $(x^2 - 3)^4 = 2$

$$x^8 - 12x^6 + 54x^4 - 108x^2 + 79 = 0$$

The equation is irreducible since,

a) $x^2 = 3 + \sqrt[4]{2}$ has the roots $\pm\sqrt{3 + \sqrt[4]{2}}$

b) If we let $y = x^2 - 3$, then the roots of $y^4 = 2$ are the 4 fourth roots of 2. Two roots are real and irrational, the other two roots are imaginary.

c) Therefore the eight roots of the equation can be represented by $x = \pm\sqrt{3 + p}$ where p is a fourth root of 2. None of these roots is rational

2) First construct $\sqrt{2}$ by obtaining the mean proportion between 2 and 1. Then construct $\sqrt[4]{2}$ by obtaining the mean propotion between $\sqrt{2}$ and 1. Third construct $\sqrt[8]{2}$ by obtaining the mean propotion between $\sqrt[4]{2}$ and 1. Fourth obtain $\sqrt[16]{2}$. Finally obtain $\sqrt{3 + \sqrt[8]{2}}$ as the bypotencese of a right triangle where legs are $\sqrt{3}$ and $\sqrt[16]{2}$.

3) a) If the coefficients of $x^n + a_1x^{n-1} + \cdots + a_n = 0$ are integers, a rational root of that equation must be an integer which divides a_n. The equation $x^3 - 1 = 0$ has a rational root $x = 1$.

   b) The divisors of 2 are ± 1 and ± 2. None of these numbers is a root of $x^3 - 2 = 0$. Therefore $x^3 - 2 = 0$ has no rational root.

4) $x^3 - 1 = (x - 1)(x^2 + x + 1) = 0$.
   Therefore the roots of $x^3 - 1 = 0$ are $x = 1$, $x = (-1 \pm \sqrt{-3})/2$.

5) If r_1 is a root of $x^3 + ax^2 + bx + c = 0$, then $(x - r_1)$ is a factor of the left hand side of the equation. Let the other factor be $x^2 + mx + n$, then $(x - r_1)(x^2 + mx + n) = 0$. The solution of the quadratic equation involves only rational operations and the extractions of real square roots (all the roots of the original equation

are real). Therefore the roots satisfy the analytic criterion for constructibility. The rational root of $x^3 - 7x^2 + 14x - 6 = 0$ is $x = 3$. $(x - 3)(x^2 - 4x + 2) = 0$. The roots of the cubic equations are 3, $2 \pm \sqrt{2}$.

6) Since $2 + \sqrt{3}$ is a root of the equation

$$(2 + \sqrt{3})^3 + a(2 + \sqrt{3})^2 + b(2 + \sqrt{3}) + c = 0.$$

Simplifying we obtain

$$(26 + 7a + 2b + c) + (15 + 4a + b)\sqrt{3} = 0.$$

Inasmuch as $\sqrt{3}$ is irrational,

$$15 + 4a + b = 0$$
$$\text{and}. \quad 26 + 7a + 2b + c = 0.$$

If $2 - \sqrt{3}$ is substituted for x in the equation we obtain

$$(26 + 7a + 2b + c) - (15 + 4a + b)\sqrt{3} = 0$$

But $26 + 7a + 2b + c = 0$
and $15 + 4a + b = 0$

$\therefore \quad 2 - \sqrt{3}$ is also a root of the equation. If the third root is represented by r,

$$(2 + \sqrt{3}) + (2 - \sqrt{3}) + r = -a$$
$$r = -a - 4$$

and r is rational.

SET III-A

1) $a = 0, b = 1; a = 0, b = -1$

SET III–B

1) $(a, b) = (c, d)$ if and only if $a = c, b = d$
   $(a, b) + (c, d) = (a + c, b + d)$
   $(a, b) \times (c, d) = (ac - bd, ad + bc)$
2) One solution is $(0, 1)$
   $(0, 1) \times (0, 1) = ([0 - 1], [0 + 0]) = (-1, 0)$
   $(-1, 0) + (1, 0) = (0, 0)$.
   Another solution is $(0, -1)$

SET III–C

1) $r_1(\cos \theta_1 + i \sin \theta_1) \cdot r_2(\cos \theta_2 + i \sin \theta_2)$
   $= r_1 r_2(\cos \theta_1 \cos \theta_2 - \sin \theta_1 \sin \theta_2 + i ([\sin \theta_1 \cos \theta_2$
   $\qquad\qquad\qquad\qquad\qquad\qquad + \cos \theta_1, \sin \theta_2])$
   $= r_1 r_2 (\cos [\theta_1 + \theta_2] + i \sin [\theta_1 + \theta_2]$.

2) $\dfrac{r_1(\cos \theta_1 + i \sin \theta_1)}{r_2(\cos \theta_2 + i \sin \theta_2)} \cdot \dfrac{(\cos \theta_2 - i \sin \theta_2)}{(\cos \theta_2 - i \sin \theta_2)}$

$\dfrac{r_1}{r_2} = \dfrac{\cos \theta_1 \cos \theta_2 + \sin \theta_1 \sin \theta_2 + i(\sin \theta_1 \cos \theta_2 - \cos \theta_1 \sin \theta_2)}{\cos^2 \theta + \sin^2 \theta}$

$\qquad = \dfrac{r_1}{r_2} (\cos [\theta_1 - \theta_2] + i \sin [\theta_1 - \theta_2])$

SET III–D

1) & 2) $(\cos x + i \sin x)^3 = \cos^3 x + 3i \cos^2 x \sin x$
$- 3 \cos x \sin^2 x - i \sin^3 x = \cos 3x + i \sin 3x$

$\cos 3x = \cos^3 x - 3 \cos x \sin^2 x = \cos^3 x$
$\qquad\qquad - 3 \cos x(1 - \cos^2 x) = 4 \cos^3 x - 3 \cos x$

$\sin 3x = 3 \cos^2 x \sin x - \sin^3 x =$
$\qquad\qquad 3(1 - \sin^2 x) \sin x - \sin^3 x = 3 \sin x - 4 \sin^3 x$

SET III–E

1) $R_1 = \cos \dfrac{2\pi}{3} + i \sin \dfrac{2\pi}{3} = -\dfrac{1}{2} + \dfrac{i\sqrt{3}}{2}$

$\qquad = \dfrac{-1 + \sqrt{-3}}{2} = x_2$

$R_2 = \cos \dfrac{4\pi}{3} + i \sin \dfrac{4\pi}{3} = -\dfrac{1}{2} - \dfrac{i\sqrt{3}}{2}$

$\qquad = \dfrac{-1 - \sqrt{3}}{2} = x_3$

$R_3 = \cos \dfrac{6\pi}{3} + i \sin \dfrac{6\pi}{3} = 1 = x_1$

SET III–F

1) $x_3 = \left(\dfrac{-1+\sqrt{-3}}{2}\right)^2 = \dfrac{-1-\sqrt{-3}}{2} = \omega^2$

$x_1 = \omega \cdot \omega^2 = \omega^3 = 1$

If we represent the negative square root of 1 by R_1, then $R_2 = 1 = R_1^2$

SET III–G

1) $R = \cos\dfrac{2\pi}{7} + i\sin\dfrac{2\pi}{7}$

$R^2 = \cos\dfrac{4\pi}{7} + i\sin\dfrac{4\pi}{7}$

$\cdot$
$\cdot$
$\cdot$

$R^7 = \cos\dfrac{14\pi}{7} + i\sin\dfrac{14\pi}{7} = 1$

2) a) $R^3 \cdot R^6 = R^9 = R^2$

b) the inverse of R^5 is R^2 since $R^5 \cdot R^2 = R^7 = 1$

c) $R^2 = \cos\dfrac{4\pi}{7} + i\sin\dfrac{4\pi}{7}$

$\dfrac{1}{R^2} = \cos\dfrac{4\pi}{7} - i\sin\dfrac{4\pi}{7}$

$R^2 + \dfrac{1}{R^2} = 2\cos\dfrac{4\pi}{7}.$

3) $r = R, a = 1$

$S = \dfrac{R^n - 1}{R - 1}.$

SET IV–A

1) If $b = 2a$, then $y^2 = bx = 2ax$ also since $x^2 = ay$, $y = x^2/a$, $x^4/a^2 = 2ax$ or $x^3 = 2a^3$

2) The graphs of $x^2 = ay$ and $y^2 = bx$ are parabolas. Although individual points of a parabola can be constructed with straight edge and compasses, the entire graph, and in particular the points of intersection of the two parabolas, cannot be constructed with straight edge and compasses.

3) Since $x^2 = ay$, $y = x^2/a$

$$xy = \frac{x^3}{a} = ab.$$

But $b = 2a$ therefore $x^3 = a^2b = 2a^3$.

SET IV–B

1) The problem is: given a line whose length is a, construct a line whose length is x, so that $x^2 = 2a^2$. $2a/x = x/a$, construct the mean proportion between $2a$ and a.

2) a) The equation is $4\pi x^2 = 2(4\pi)$. Therefore $x^2 = 2$ and the construction is possible.

b) $\frac{4\pi x^3}{3} = \frac{8\pi}{3}$, $x^3 = 2$, and the construction is impossible.

SET V–A

1) OD is the cos A.

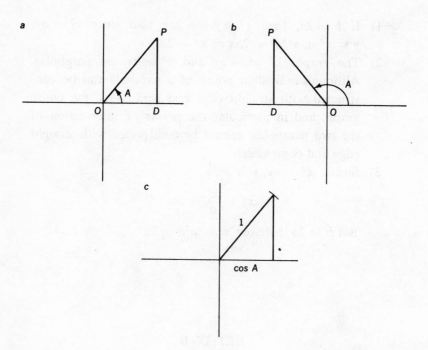

2) a) If $3\theta = 90°$, $\theta = 30°$ and the equation becomes $x^3 - 3x = 0$, $(x = 2\cos\theta)$. Since this equation has a rational root $(x = 0)$, it is a reducible cubic and its root can be constructed i.e. an angle of 90° can be trisected.

b) For $3\theta = 120°$ the equation is the irreducible cubic $x^3 - 3x + 1 = 0$. An angle of 120° cannot be trisected.

c) For $3\theta = 180°$, the equation is the reducible cubic, $x^3 - 3x + 2 = 0$ (One root is $x = 1$. An angle of 180° can be trisected).

SET V–B

1) Let angle D contain $y°$. Draw OE then $\angle$ DOE $= y°$ and $\angle$ BEO $= 2y°$. Also $\angle$ B $= 2y°$ then $\angle$ AOB $= 3y°$ or $\angle$ D $= 1/3$ of $\angle$ AOB.

2) To construct line DEB so that DE is equal to the radius of the circle would require placing two marks on the straight edge (which should be unmarked).

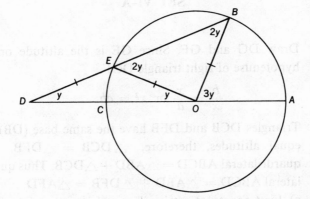

SET V–C

1) Use the formula for the sum of an infinite geometric progression with $a = \frac{1}{4}$ and $s = \frac{1}{4}$, $S = \frac{1}{4}/(1 - \frac{1}{4})$ $= \frac{1}{4}/\frac{3}{4} = \frac{1}{3}$

2) Only a finite number of operations with straight edge and compasses are permitted. This method requires that the given angle be bisected an infinite number of times.

SET VI–A

1) Draw DG and GF. Since GE is the altitude on the hypotenuse of right triangle DGF

$$\frac{b}{x} = \frac{x}{a} \text{ or } x^2 = ab.$$

2) Triangles DCB and DFB have the same base (DB) and equal altitudes, therefore, $\triangle DCB = \triangle DFB$. Also quardrilateral $ABCD = \triangle ABD + \triangle DCB$. Thus quadrilateral $ABCD = \triangle ABD + \triangle DFB = \triangle AFD$.

3) a) First construct a triangle equal in area to the given quadrilateral, then construct x to be the mean proportion between $\frac{1}{2} b$ and h, when b and h are any base and the corresponding altitude of the triangle which is equal to the quadrilateral.

b) Let ABCDE be the given pentagon. Draw diagonals DB and DA. Construct CG // DB and EF // DA.

(F and G lie on AB extended.) Draw DF and DG. Now prove that $\triangle$DFG is equal to pentagon ABCDE. Construct a square equal to $\triangle$DFG. This method can be extended to construct a square equal in area to any given polygon.

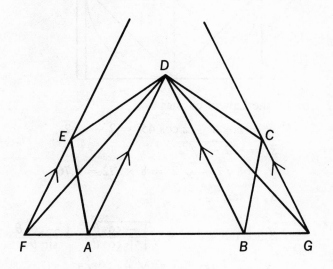

SET VI–B

1) a) Let s_n = length of side of inscribed regular polygon of n sides; S_n = length of side of regular circumscribed polygon; p_n = perimeter of inscribed polygon and P_n = perimeter circumscribed regular polygon

$$s_4 = \sqrt{2}, S_4 = 2.$$
$$p_4 = 4\sqrt{2} = 4(1.414) = 5.66$$
$$P_4 = 8$$
$$5.66 \leq 2\pi \leq 8$$
$$2.83 \leq \pi \leq 4$$

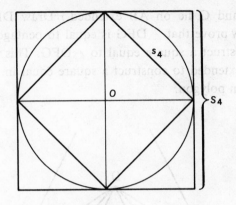

b) By the Law of cosines

$$s_8^2 = 1 + 1 - 2 \cos 45° = 2 - \sqrt{2}$$
$$s_8 = \sqrt{2 - \sqrt{2}}$$
$$p_8 = 8\sqrt{2 - \sqrt{2}} = 8 \times .72 = 5.76$$
$$S_8 = \tan \frac{45°}{2}$$

Since $\tan \frac{1}{2}\theta = \pm\sqrt{\dfrac{1 - \cos\theta}{1 + \cos\theta}} = \dfrac{1 - \cos\theta}{\sin\theta}$

$$\tan \frac{1}{2}45° = \frac{2 - \sqrt{2}}{\sqrt{2}} = \sqrt{2} - 1$$

$$P_8 = 16(\sqrt{2} - 1) = 16(.414) = 6.62$$
$$2.88 \leq \pi \leq 3.31$$

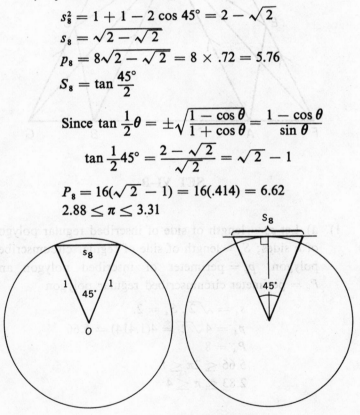

2) $3\frac{1}{7} = 3.\overline{142857}$

3) Let $x = 3.\overline{14}$

$$100x = 314.\overline{14}$$
$$99x = 311$$
$$x = \frac{311}{99}$$

SET VI–C

1) Draw OC, CD, DA, CA

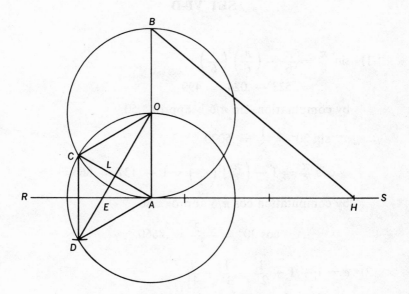

OCDA is a rhombus

$\angle COA = 60°, \angle CAE = 30°$

$OD \perp CA$

$\therefore \quad EA = \dfrac{\sqrt{3}}{3}$

$AH = 3 - \dfrac{\sqrt{3}}{3}, \quad AB = 2$

By the Pythagorian theorem

$$BH = \sqrt{\tfrac{40}{3} - 2\sqrt{3}}$$

2) $BH = 3.14153$, $\pi = 3.14159$ the error is approximately .00006.

SET VI–D

1) $\sin \dfrac{\pi}{6} = \dfrac{\pi}{6} - \left(\dfrac{\pi}{6}\right)^3\left(\dfrac{1}{6}\right)$

$\qquad\qquad = .523 - .024 = .499$

by computation, $\sin \pi/6$ is approx. .50,

$\qquad \sin 30° = \dfrac{1}{2} = .5000$

$\qquad \cos \dfrac{\pi}{6} = 1 - \left(\dfrac{\pi}{6}\right)^2\left(\dfrac{1}{2}\right) = 1 - .137 = .863$

by computation $\cos \pi/6$ approx. .86

$\qquad \cos 30° = \dfrac{\sqrt{3}}{2} = .8660.$

2) $e = 1 + 1 + \dfrac{1}{2!} + \dfrac{1}{3!} + \dfrac{1}{4!}$

$e = 2 + .5 + .167 + .042 = 2.709$

e is approx. 2.71

SET VI–E

$$x = \sqrt{\tfrac{40}{3} - 2\sqrt{3}}$$
$$x^2 = \tfrac{40}{3} - 2\sqrt{3}$$
$$3x^2 - 40 = -6\sqrt{3}$$
$$9x^4 - 240x^2 + 1492 = 0$$

SET VII–A

1) a) Construct a square by joining the ends of two perpendecular diameters. Bisecting the sides of a regular polygon of n sides will give the vertices of a regular polygon of $2n$ sides.

   b) Construct a regular hexagon, using the radius of a circle to divide the circle into 6 equal parts. Join the alternate vertices of the hexagon.

   c) See part a).

2) a) see (1) b.

   b) see (1) b.

   c) see (1) a.

3) For $k = 2$ and $l = -1$, $3k + 5l = 1$. We can then construct central angles of $2\pi/3$ and $2\pi/5$, also we can construct angles of $2\pi(\tfrac{1}{3})$ and $2\pi(\tfrac{2}{5})$, we can then construct an angle of $2\pi(\tfrac{2}{5} - \tfrac{1}{3}) = 2\pi/15$, which is a central angle of a regular polygon of 15 sides.

4) a) Since 3 is not relatively prime to itself, the theorem gives us no information about a regular polygon of 3×3 or 9 sides. However, the central angle of a regular polygon of 9 sides is $40°$, and we have already shown

that such an angle cannot be trisected.

b) The statement is false. A regular polygon of 5 sides can be constructed and a regular polygon of 4 sides can be constructed. The integers 4 and 5 are relatively prime, but a regular polygon of 9 sides cannot be constructed.

SET VII–B

1) Construct a circle of unit radius (any convenient radius). Divide the radius into a mean and extreme ratio; i.e., construct x so that $x = (\sqrt{5} - 1)/2$. Use x as a side of a regular decagon, finally join the alternate vertices.

2) $\angle 1 = \angle 2$

$\angle 1 = \angle 3, \angle 2 = \angle 4$

$\therefore \quad \angle 3 = \angle 4$

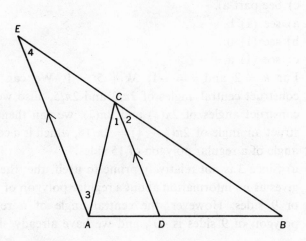

and $AC = EC$.

But $\dfrac{EC}{CB} = \dfrac{AD}{DB}$

$\therefore \quad \dfrac{AC}{CB} = \dfrac{AD}{DB}$.

SET VII–C

1) Expand $(x - r_1)(x - r_2)(x - r_3)$

then

$$x^3 - (r_1 + r_2 + r_3)x^2 + (r_1r_2 + r_1r_3 + r_2r_3)x$$
$$- r_1r_2r_3 \equiv x^3 + a_1x^2 + a_2x + a_3.$$

When two polynomials are identically equal, the coefficient of corresponding terms are equal (an equation of degree n cannot have more than n roots; whereas in an identity any permissible value of x will make the equations true). Equating the coefficients of like terms gives the desired relation between the roots and the coefficients.

2) Use the same procedure. This procedure can be used for an equation of degree n.

SET VII–D

1) Since $y_1 + y_2 = -1$ and $y_1 \cdot y_2 = -1$, y_1 and y_2 satisfy the equation $y^2 + y - 1 = 0$.

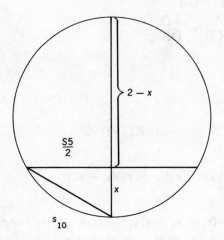

2) Since OC $= 1/2$, OB $= 1$, BC $= \sqrt{5}/2$, CD $= \sqrt{5}/2$ and OD $= (\sqrt{5} - 1)/2 = s_{10}$.

To prove $s_5 =$ BD, we have $s_5^2/4 + x^2 = s_{10}^2$ also $2x - x^2 = s_5^2/4$ $\therefore$ $x = s_{10}^2/2$

and $\quad s_5^2 + s_{10}^4 = 4s_{10}^2$

$$s_5^2 = s_{10}^2 (4 - s_{10}^2)$$

$$s_{10} = \frac{-1 + \sqrt{5}}{2}, \quad s_{10}^2 = \frac{3 - \sqrt{5}}{2}$$

then

$$s_5^2 - 1 = \frac{3 - \sqrt{5}}{2}\left(\frac{5 + \sqrt{5}}{2}\right) - 1 = \frac{3 - \sqrt{5}}{2} = s_{10}^2$$

$$\therefore \quad s_5^2 = 1 + s_{10}^2$$

$$\text{or} \quad s_5 = \text{BD}.$$

SET VII–E

1) $R^3 = \cos\dfrac{2\pi}{3} + i\sin\dfrac{2\pi}{3} = \omega$

$R^6 = \cos\dfrac{4\pi}{3} + i\sin\dfrac{4\pi}{3} = \omega^2$

where ω is a cube root of unity

$$\therefore \quad R^3 = \frac{-1 + \sqrt{-3}}{2}, \; R^6 = \frac{-1 - \sqrt{-3}}{2}$$

2) $y_1y_2 + y_1y_3 + y_2y_3 = 3(R^3 + R^6) + (R + R^2 + R^4$
$$+ R^5 + R^7 + R^8)$$
$$= 3(-1) + 0$$

since R^3, R^6 are the roots of $x^2 + x + 1 = 0$

$y_1y_2y_3 = (R^3 + R^6) + (R + R^2 + R^4 + R^5 + R^7$
$$+ R^8)$$
$$= -1 + 0 = -1$$

3) Trisecting an angle of 120° is equivalent to constructing an angle of 40°. A regular polygon of 9 sides subtends an angle of 40° at the center of the circumscribed circle.

SET VII–F

1) Perform the actual division.

SET VII–G

1) If $g = 2$, we obtain the following sequence R, R², R⁴,

—

R^8, R^{16}, $R^{32}(=R^{15})$, $R^{30}(=R^{13})$, $R^{26}(=R^9)$, $R^{18}=R$, and the sequence will continue R^2, R^4, etc. Not all the roots of the cyclotonic equation are in the sequences.

2) Each term is the cube of the preceeding term. When the exponent of R becomes greater than 16, subtract 17 from the exponent. This is equivalent to dividing the term by 1 since $R^{17}=1$.

SET VII–H

1) Since $y^2 + y - 4 = 0$ and $y_1 > 0$

$$y_1 = \frac{-1 + \sqrt{17}}{2} \quad \text{and} \quad y_2 = \frac{-1 - \sqrt{17}}{2}$$

also

$$z^2 - y_1 z - 1 = 0 \quad \text{and} \quad z_1 > 0$$

Therefore

$$z_1 = \frac{y_1 + \sqrt{y_1^2 + 4}}{2} = \frac{1}{2}y_1 + \sqrt{1 + \frac{1}{4}y_1^2}$$

Finally,

$$w^2 - y_2 w - 1 = 0 \quad \text{and} \quad w_1 > 0$$

Therefore

$$w_1 = \frac{1}{2}y_2 + \sqrt{1 + \frac{1}{4}y_2^2}$$

2) a) $\overline{OE}^2 = \overline{AE}^2 + \overline{OA}^2 = \frac{1}{16} + 1$

$$\overline{OE} = \frac{1}{4}\sqrt{17}$$

b) $\overline{AF} = \overline{EF} - \overline{EA} = \overline{OE} - \overline{EA} = \frac{1}{4}\sqrt{17} - \frac{1}{4} = \frac{1}{2}y_1$

c) Similarly

$$\overline{AF'} + \overline{EF'} + \overline{AE} = \frac{1}{4}\sqrt{17} + \frac{1}{4} = -\frac{1}{2}y_2$$

d) $\overline{OF}^2 = \overline{OA}^2 + \overline{AF}^2 = 1 + \left(\frac{1}{2}y_1\right)^2$

$$\overline{OF} = \sqrt{1 + \frac{1}{4}y_1^2}$$

$$\overline{OF'}^2 = \overline{OA}^2 + \overline{AF'}^2 = 1 + \left(\frac{1}{2}y_2\right)^2$$

$$\overline{OF'} = \sqrt{1 + \frac{1}{4}y_2^2}$$

e) $\overline{AH} = \overline{AF} + \overline{FH} = \frac{1}{2}y_1 + \overline{OF}$

$$= \frac{1}{2}y_1 + \sqrt{1 + \frac{1}{4}y_1^2} = z_1$$

f) $\overline{AH'} = \overline{F'H'} - \overline{F'A} = \overline{F'O} - \left(-\frac{1}{2}y_2\right)$

$$= \sqrt{1 + \frac{1}{4}y_2^2} + \frac{1}{2}y_2 = w_1$$

3) The equation of the circle is

$$\left(x - \frac{z_1}{2}\right)^2 + \left(y - \frac{w_1 + 1}{2}\right)^2 = \left(\frac{z_1}{2}\right)^2 + \left(\frac{w_1 - 1}{2}\right)^2$$

therefore the abscissas of the points G and F can be obtained from the equation

$$\left(x - \frac{z_1}{2}\right)^2 + \left(\frac{w_1 + 1}{2}\right)^2 = \left(\frac{z_1}{2}\right)^2 + \left(\frac{w_1 - 1}{2}\right)^2$$

or $x^2 - z_1 x + w_1 = 0$

But v_1 and v_2 satisfy the equation

$$v^2 - z_1 v + w_1 = 0 \text{ and } v_1 > v_2 > 0.$$

therefore

$$\text{OF} = v_1\left(= \frac{z_1 + \sqrt{z_1^2 - 4w_1}}{2}\right)$$

Also

$$v_1 = R + \frac{1}{R} = \cos\frac{2\pi}{17}$$

Consequently $OF = 2\cos\frac{2\pi}{17}$.

4) $\overline{OM} = \frac{1}{2}\,OE = \cos\frac{2\pi}{17}$

since the radius of the cirlce is unity,

$$\angle\,MOP = \frac{2\pi}{17}.$$

5) Follow the procedure in the text.

SET VII–K

$n =$	2^m	3.2^m	5.2^m	17.2^m	15.2^m	51.2^m	85.2^m
	4	3	5	17	15	51	85
	8	6	10	34	30		
	16	12	20	68	60		
	32	24	40				
	64	48	80				
		96					
Totals	5	6	5	3	3	1	1

There are 24 regular polygons with number of sides ≤ 100 which can be constructed.

SET VII–L

1) $2^{10} = 1024$

   $\log 2^{2^{10}} = 1024 \log 2 = 1024 \,(.301) = 308.224.$

   Therefore $2^{2^{10}} + 1$ has 309 digits.

2) We can construct regular polygons of 12 sides and 15 sides. Therefore we can construct angles of 30° and 24°. The difference of these two angles can he constructed. Finally the angle of 6° can be bisected to give an angle of 3°. We cannot construct an angle of 2°. For, if this were possible we could then construct an angle of 1° and any integral multiple of 1°; in particular an angle of 40°. But a regular polygon of 9 sides cannot be constructed. Therefore the smallest integral number of degrees an angle can have and still be constructible is 3.

SET VIII–A

1) If $AB = AC$, $\angle B = \angle C$.

   Then $\triangle ABD \cong \triangle ACE$

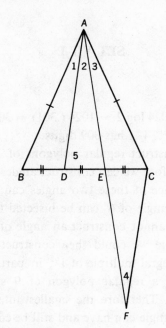

and $\angle 1 = \angle 3$.

Suppose $\angle 2 = \angle 3$.

Extend AE its own length to F and draw CF.

$\triangle$ADE $\cong$ $\triangle$FCE, and $\angle 2 = \angle 4$

∴ $\angle 3 = \angle 4$ and AC = CF.

Also, since AD = CF, AD = AC

∴ $\angle 5 = \angle C = \angle B$

    but $\angle 5 > \angle B$

∴   The assumption that $\angle 2 = \angle 3$ is false.

A CATALOGUE OF
SELECTED DOVER BOOKS
IN ALL FIELDS OF INTEREST

A CATALOGUE OF SELECTED DOVER
BOOKS IN ALL FIELDS OF INTEREST

RACKHAM'S COLOR ILLUSTRATIONS FOR WAGNER'S RING. Rackham's finest mature work—all 64 full-color watercolors in a faithful and lush interpretation of the *Ring*. Full-sized plates on coated stock of the paintings used by opera companies for authentic staging of Wagner. Captions aid in following complete Ring cycle. Introduction. 64 illustrations plus vignettes. 72pp. 8⅝ x 11¼.   23779-6 Pa. $6.00

CONTEMPORARY POLISH POSTERS IN FULL COLOR, edited by Joseph Czestochowski. 46 full-color examples of brilliant school of Polish graphic design, selected from world's first museum (near Warsaw) dedicated to poster art. Posters on circuses, films, plays, concerts all show cosmopolitan influences, free imagination. Introduction. 48pp. 9⅜ x 12¼.
23780-X Pa. $6.00

GRAPHIC WORKS OF EDVARD MUNCH, Edvard Munch. 90 haunting, evocative prints by first major Expressionist artist and one of the greatest graphic artists of his time: *The Scream, Anxiety, Death Chamber, The Kiss, Madonna*, etc. Introduction by Alfred Werner. 90pp. 9 x 12.
23765-6 Pa. $5.00

THE GOLDEN AGE OF THE POSTER, Hayward and Blanche Cirker. 70 extraordinary posters in full colors, from Maitres de l'Affiche, Mucha, Lautrec, Bradley, Cheret, Beardsley, many others. Total of 78pp. 9⅜ x 12¼.   22753-7 Pa. $5.95

THE NOTEBOOKS OF LEONARDO DA VINCI, edited by J. P. Richter. Extracts from manuscripts reveal great genius; on painting, sculpture, anatomy, sciences, geography, etc. Both Italian and English. 186 ms. pages reproduced, plus 500 additional drawings, including studies for *Last Supper*, Sforza monument, etc. 860pp. 7⅞ x 10¾. (Available in U.S. only)
22572-0, 22573-9 Pa., Two-vol. set $15.90

THE CODEX NUTTALL, as first edited by Zelia Nuttall. Only inexpensive edition, in full color, of a pre-Columbian Mexican (Mixtec) book. 88 color plates show kings, gods, heroes, temples, sacrifices. New explanatory, historical introduction by Arthur G. Miller. 96pp. 11⅜ x 8½. (Available in U.S. only)   23168-2 Pa. $7.95

UNE SEMAINE DE BONTÉ, A SURREALISTIC NOVEL IN COLLAGE, Max Ernst. Masterpiece created out of 19th-century periodical illustrations, explores worlds of terror and surprise. Some consider this Ernst's greatest work. 208pp. 8⅛ x 11.   23252-2 Pa. $5.00

DRAWINGS OF WILLIAM BLAKE, William Blake. 92 plates from Book of Job, *Divine Comedy, Paradise Lost,* visionary heads, mythological figures, Laocoon, etc. Selection, introduction, commentary by Sir Geoffrey Keynes. 178pp. 8⅛ x 11.   22303-5 Pa. $4.00

ENGRAVINGS OF HOGARTH, William Hogarth. 101 of Hogarth's greatest works: *Rake's Progress, Harlot's Progress, Illustrations for Hudibras, Before and After, Beer Street and Gin Lane,* many more. Full commentary. 256pp. 11 x 13¾.   22479-1 Pa. $12.95

DAUMIER: 120 GREAT LITHOGRAPHS, Honore Daumier. Wide-ranging collection of lithographs by the greatest caricaturist of the 19th century. Concentrates on eternally popular series on lawyers, on married life, on liberated women, etc. Selection, introduction, and notes on plates by Charles F. Ramus. Total of 158pp. 9⅜ x 12¼.   23512-2 Pa. $5.50

DRAWINGS OF MUCHA, Alphonse Maria Mucha. Work reveals drafts-man of highest caliber: studies for famous posters and paintings, render-ings for book illustrations and ads, etc. 70 works, 9 in color; including 6 items not drawings. Introduction. List of illustrations. 72pp. 9⅜ x 12¼. (Available in U.S. only)   23672-2 Pa. $4.00

GIOVANNI BATTISTA PIRANESI: DRAWINGS IN THE PIERPONT MORGAN LIBRARY, Giovanni Battista Piranesi. For first time ever all of Morgan Library's collection, world's largest. 167 illustrations of rare Piranesi drawings—archeological, architectural, decorative and visionary. Essay, detailed list of drawings, chronology, captions. Edited by Felice Stampfle. 144pp. 9⅜ x 12¼.   23714-1 Pa. $7.50

NEW YORK ETCHINGS (1905-1949), John Sloan. All of important American artist's N.Y. life etchings. 67 works include some of his best art; also lively historical record—Greenwich Village, tenement scenes. Edited by Sloan's widow. Introduction and captions. 79pp. 8⅜ x 11¼.
                                                                        23651-X Pa. $4.00

CHINESE PAINTING AND CALLIGRAPHY: A PICTORIAL SURVEY, Wan-go Weng. 69 fine examples from John M. Crawford's matchless private collection: landscapes, birds, flowers, human figures, etc., plus calligraphy. Every basic form included: hanging scrolls, handscrolls, album leaves, fans, etc. 109 illustrations. Introduction. Captions. 192pp. 8⅞ x 11¾.
                                                                        23707-9 Pa. $7.95

DRAWINGS OF REMBRANDT, edited by Seymour Slive. Updated Lipp-mann, Hofstede de Groot edition, with definitive scholarly apparatus. All portraits, biblical sketches, landscapes, nudes, Oriental figures, classical studies, together with selection of work by followers. 550 illustrations. Total of 630pp. 9⅛ x 12¼.   21485-0, 21486-9 Pa., Two-vol. set $15.00

THE DISASTERS OF WAR, Francisco Goya. 83 etchings record horrors of Napoleonic wars in Spain and war in general. Reprint of 1st edition, plus 3 additional plates. Introduction by Philip Hofer. 97pp. 9⅜ x 8¼.
                                                                        21872-4 Pa. $3.75

THE EARLY WORK OF AUBREY BEARDSLEY, Aubrey Beardsley. 157 plates, 2 in color: *Manon Lescaut, Madame Bovary, Morte Darthur, Salome,* other. Introduction by H. Marillier. 182pp. 8⅛ x 11. 21816-3 Pa. $4.50

THE LATER WORK OF AUBREY BEARDSLEY, Aubrey Beardsley. Exotic masterpieces of full maturity: *Venus and Tannhauser, Lysistrata, Rape of the Lock, Volpone,* Savoy material, etc. 174 plates, 2 in color. 186pp. 8⅛ x 11. 21817-1 Pa. $4.50

THOMAS NAST'S CHRISTMAS DRAWINGS, Thomas Nast. Almost all Christmas drawings by creator of image of Santa Claus as we know it, and one of America's foremost illustrators and political cartoonists. 66 illustrations. 3 illustrations in color on covers. 96pp. 8⅜ x 11¼. 23660-9 Pa. $3.50

THE DORÉ ILLUSTRATIONS FOR DANTE'S DIVINE COMEDY, Gustave Doré. All 135 plates from Inferno, Purgatory, Paradise; fantastic tortures, infernal landscapes, celestial wonders. Each plate with appropriate (translated) verses. 141pp. 9 x 12. 23231-X Pa. $4.50

DORÉ'S ILLUSTRATIONS FOR RABELAIS, Gustave Doré. 252 striking illustrations of *Gargantua and Pantagruel* books by foremost 19th-century illustrator. Including 60 plates, 192 delightful smaller illustrations. 153pp. 9 x 12. 23656-0 Pa. $5.00

LONDON: A PILGRIMAGE, Gustave Doré, Blanchard Jerrold. Squalor, riches, misery, beauty of mid-Victorian metropolis; 55 wonderful plates, 125 other illustrations, full social, cultural text by Jerrold. 191pp. of text. 9⅜ x 12¼. 22306-X Pa. $7.00

THE RIME OF THE ANCIENT MARINER, Gustave Doré, S. T. Coleridge. Dore's finest work, 34 plates capture moods, subtleties of poem. Full text. Introduction by Millicent Rose. 77pp. 9¼ x 12. 22305-1 Pa. $3.50

THE DORE BIBLE ILLUSTRATIONS, Gustave Doré. All wonderful, detailed plates: Adam and Eve, Flood, Babylon, Life of Jesus, etc. Brief King James text with each plate. Introduction by Millicent Rose. 241 plates. 241pp. 9 x 12. 23004-X Pa. $6.00

THE COMPLETE ENGRAVINGS, ETCHINGS AND DRYPOINTS OF ALBRECHT DURER. "Knight, Death and Devil"; "Melencolia," and more—all Dürer's known works in all three media, including 6 works formerly attributed to him. 120 plates. 235pp. 8⅜ x 11¼. 22851-7 Pa. $6.50

MAXIMILIAN'S TRIUMPHAL ARCH, Albrecht Dürer and others. Incredible monument of woodcut art: 8 foot high elaborate arch—heraldic figures, humans, battle scenes, fantastic elements—that you can assemble yourself. Printed on one side, layout for assembly. 143pp. 11 x 16. 21451-6 Pa. $5.00

THE COMPLETE WOODCUTS OF ALBRECHT DURER, edited by Dr. W. Kurth. 346 in all: "Old Testament," "St. Jerome," "Passion," "Life of Virgin," Apocalypse," many others. Introduction by Campbell Dodgson. 285pp. 8½ x 12¼.   21097-9 Pa. $7.50

DRAWINGS OF ALBRECHT DURER, edited by Heinrich Wolfflin. 81 plates show development from youth to full style. Many favorites; many new. Introduction by Alfred Werner. 96pp. 8⅛ x 11.   22352-3 Pa. $5.00

THE HUMAN FIGURE, Albrecht Dürer. Experiments in various techniques—stereometric, progressive proportional, and others. Also life studies that rank among finest ever done. Complete reprinting of *Dresden Sketchbook*. 170 plates. 355pp. 8⅜ x 11¼.   21042-1 Pa. $7.95

OF THE JUST SHAPING OF LETTERS, Albrecht Dürer. Renaissance artist explains design of Roman majuscules by geometry, also Gothic lower and capitals. Grolier Club edition. 43pp. 7⅞ x 10¾   21306-4 Pa. $3.00

TEN BOOKS ON ARCHITECTURE, Vitruvius. The most important book ever written on architecture. Early Roman aesthetics, technology, classical orders, site selection, all other aspects. Stands behind everything since. Morgan translation. 331pp. 5⅜ x 8½.   20645-9 Pa. $4.50

THE FOUR BOOKS OF ARCHITECTURE, Andrea Palladio. 16th-century classic responsible for Palladian movement and style. Covers classical architectural remains, Renaissance revivals, classical orders, etc. 1738 Ware English edition. Introduction by A. Placzek. 216 plates. 110pp. of text. 9½ x 12¾.   21308-0 Pa. $10.00

HORIZONS, Norman Bel Geddes. Great industrialist stage designer, "father of streamlining," on application of aesthetics to transportation, amusement, architecture, etc. 1932 prophetic account; function, theory, specific projects. 222 illustrations. 312pp. 7⅞ x 10¾.   23514-9 Pa. $6.95

FRANK LLOYD WRIGHT'S FALLINGWATER, Donald Hoffmann. Full, illustrated story of conception and building of Wright's masterwork at Bear Run, Pa. 100 photographs of site, construction, and details of completed structure. 112pp. 9¼ x 10.   23671-4 Pa. **$5.50**

THE ELEMENTS OF DRAWING, John Ruskin. Timeless classic by great Viltorian; starts with basic ideas, works through more difficult. Many practical exercises. 48 illustrations. Introduction by Lawrence Campbell. 228pp. 5⅜ x 8½.   22730-8 Pa. $3.75

GIST OF ART, John Sloan. Greatest modern American teacher, Art Students League, offers innumerable hints, instructions, guided comments to help you in painting. Not a formal course. 46 illustrations. Introduction by Helen Sloan. 200pp. 5⅜ x 8½.   23435-5 Pa. **$4.00**

THE ANATOMY OF THE HORSE, George Stubbs. Often considered the great masterpiece of animal anatomy. Full reproduction of 1766 edition, plus prospectus; original text and modernized text. 36 plates. Introduction by Eleanor Garvey. 121pp. 11 x 14¾. 23402-9 Pa. $6.00

BRIDGMAN'S LIFE DRAWING, George B. Bridgman. More than 500 illustrative drawings and text teach you to abstract the body into its major masses, use light and shade, proportion; as well as specific areas of anatomy, of which Bridgman is master. 192pp. 6½ x 9¼. (Available in U.S. only) 22710-3 Pa. $3.50

ART NOUVEAU DESIGNS IN COLOR, Alphonse Mucha, Maurice Verneuil, Georges Auriol. Full-color reproduction of *Combinaisons ornementales* (c. 1900) by Art Nouveau masters. Floral, animal, geometric, interlacings, swashes—borders, frames, spots—all incredibly beautiful. 60 plates, hundreds of designs. 9⅜ x 8-1/16. 22885-1 Pa. $4.00

FULL-COLOR FLORAL DESIGNS IN THE ART NOUVEAU STYLE, E. A. Seguy. 166 motifs, on 40 plates, from *Les fleurs et leurs applications decoratives* (1902): borders, circular designs, repeats, allovers, "spots." All in authentic Art Nouveau colors. 48pp. 9⅜ x 12¼. 23439-8 Pa. $5.00

A DIDEROT PICTORIAL ENCYCLOPEDIA OF TRADES AND INDUSTRY, edited by Charles C. Gillispie. 485 most interesting plates from the great French Encyclopedia of the 18th century show hundreds of working figures, artifacts, process, land and cityscapes; glassmaking, papermaking, metal extraction, construction, weaving, making furniture, clothing, wigs, dozens of other activities. Plates fully explained. 920pp. 9 x 12. 22284-5, 22285-3 Clothbd., Two-vol. set $40.00

HANDBOOK OF EARLY ADVERTISING ART, Clarence P. Hornung. Largest collection of copyright-free early and antique advertising art ever compiled. Over 6,000 illustrations, from Franklin's time to the 1890's for special effects, novelty. Valuable source, almost inexhaustible.
Pictorial Volume. Agriculture, the zodiac, animals, autos, birds, Christmas, fire engines, flowers, trees, musical instruments, ships, games and sports, much more. Arranged by subject matter and use. 237 plates. 288pp. 9 x 12. 20122-8 Clothbd. $14.50

Typographical Volume. Roman and Gothic faces ranging from 10 point to 300 point, "Barnum," German and Old English faces, script, logotypes, scrolls and flourishes, 1115 ornamental initials, 67 complete alphabets, more. 310 plates. 320pp. 9 x 12. 20123-6 Clothbd. $15.00

CALLIGRAPHY (CALLIGRAPHIA LATINA), J. G. Schwandner. High point of 18th-century ornamental calligraphy. Very ornate initials, scrolls, borders, cherubs, birds, lettered examples. 172pp. 9 x 13. 20475-8 Pa. $7.00

ART FORMS IN NATURE, Ernst Haeckel. Multitude of strangely beautiful natural forms: Radiolaria, Foraminifera, jellyfishes, fungi, turtles, bats, etc. All 100 plates of the 19th-century evolutionist's *Kunstformen der Natur* (1904). 100pp. 9⅜ x 12¼. 22987-4 Pa. $5.00

CHILDREN: A PICTORIAL ARCHIVE FROM NINETEENTH-CENTURY SOURCES, edited by Carol Belanger Grafton. 242 rare, copyright-free wood engravings for artists and designers. Widest such selection available. All illustrations in line. 119pp. 8⅜ x 11¼. 23694-3 Pa. $3.50

WOMEN: A PICTORIAL ARCHIVE FROM NINETEENTH-CENTURY SOURCES, edited by Jim Harter. 391 copyright-free wood engravings for artists and designers selected from rare periodicals. Most extensive such collection available. All illustrations in line. 128pp. 9 x 12. 23703-6 Pa. $4.50

ARABIC ART IN COLOR, Prisse d'Avennes. From the greatest ornamentalists of all time—50 plates in color, rarely seen outside the Near East, rich in suggestion and stimulus. Includes 4 plates on covers. 46pp. 9⅜ x 12¼. 23658-7 Pa. $6.00

AUTHENTIC ALGERIAN CARPET DESIGNS AND MOTIFS, edited by June Beveridge. Algerian carpets are world famous. Dozens of geometrical motifs are charted on grids, color-coded, for weavers, needleworkers, craftsmen, designers. 53 illustrations plus 4 in color. 48pp. 8¼ x 11. (Available in U.S. only) 23650-1 Pa. $1.75

DICTIONARY OF AMERICAN PORTRAITS, edited by Hayward and Blanche Cirker. 4000 important Americans, earliest times to 1905, mostly in clear line. Politicians, writers, soldiers, scientists, inventors, industrialists, Indians, Blacks, women, outlaws, etc. Identificatory information. 756pp. 9¼ x 12¾. 21823-6 Clothbd. $40.00

HOW THE OTHER HALF LIVES, Jacob A. Riis. Journalistic record of filth, degradation, upward drive in New York immigrant slums, shops, around 1900. New edition includes 100 original Riis photos, monuments of early photography. 233pp. 10 x 7⅞. 22012-5 Pa. $7.00

NEW YORK IN THE THIRTIES, Berenice Abbott. Noted photographer's fascinating study of city shows new buildings that have become famous and old sights that have disappeared forever. Insightful commentary. 97 photographs. 97pp. 11⅜ x 10. 22967-X Pa. $5.00

MEN AT WORK, Lewis W. Hine. Famous photographic studies of construction workers, railroad men, factory workers and coal miners. New supplement of 18 photos on Empire State building construction. New introduction by Jonathan L. Doherty. Total of 69 photos. 63pp. 8 x 10¾. 23475-4 Pa. $3.00

THE DEPRESSION YEARS AS PHOTOGRAPHED BY ARTHUR ROTH-STEIN, Arthur Rothstein. First collection devoted entirely to the work of outstanding 1930s photographer: famous dust storm photo, ragged children, unemployed, etc. 120 photographs. Captions. 119pp. 9¼ x 10¾.
23590-4 Pa. $5.00

CAMERA WORK: A PICTORIAL GUIDE, Alfred Stieglitz. All 559 illustrations and plates from the most important periodical in the history of art photography, Camera Work (1903-17). Presented four to a page, reduced in size but still clear, in strict chronological order, with complete captions. Three indexes. Glossary. Bibliography. 176pp. 8⅜ x 11¼.
23591-2 Pa. $6.95

ALVIN LANGDON COBURN, PHOTOGRAPHER, Alvin L. Coburn. Revealing autobiography by one of greatest photographers of 20th century gives insider's version of Photo-Secession, plus comments on his own work. 77 photographs by Coburn. Edited by Helmut and Alison Gernsheim. 160pp. 8⅛ x 11.
23685-4 Pa. $6.00

NEW YORK IN THE FORTIES, Andreas Feininger. 162 brilliant photographs by the well-known photographer, formerly with Life magazine, show commuters, shoppers, Times Square at night, Harlem nightclub, Lower East Side, etc. Introduction and full captions by John von Hartz. 181pp. 9¼ x 10¾.
23585-8 Pa. $6.00

GREAT NEWS PHOTOS AND THE STORIES BEHIND THEM, John Faber. Dramatic volume of 140 great news photos, 1855 through 1976, and revealing stories behind them, with both historical and technical information. Hindenburg disaster, shooting of Oswald, nomination of Jimmy Carter, etc. 160pp. 8¼ x 11.
23667-6 Pa. $5.00

THE ART OF THE CINEMATOGRAPHER, Leonard Maltin. Survey of American cinematography history and anecdotal interviews with 5 masters—Arthur Miller, Hal Mohr, Hal Rosson, Lucien Ballard, and Conrad Hall. Very large selection of behind-the-scenes production photos. 105 photographs. Filmographies. Index. Originally Behind the Camera. 144pp. 8¼ x 11.
23686-2 Pa. $5.00

DESIGNS FOR THE THREE-CORNERED HAT (LE TRICORNE), Pablo Picasso. 32 fabulously rare drawings—including 31 color illustrations of costumes and accessories—for 1919 production of famous ballet. Edited by Parmenia Migel, who has written new introduction. 48pp. 9⅜ x 12¼. (Available in U.S. only)
23709-5 Pa. $5.00

NOTES OF A FILM DIRECTOR, Sergei Eisenstein. Greatest Russian filmmaker explains montage, making of Alexander Nevsky, aesthetics; comments on self, associates, great rivals (Chaplin), similar material. 78 illustrations. 240pp. 5⅜ x 8½.
22392-2 Pa. $4.50

HOLLYWOOD GLAMOUR PORTRAITS, edited by John Kobal. 145 photos capture the stars from 1926-49, the high point in portrait photography. Gable, Harlow, Bogart, Bacall, Hedy Lamarr, Marlene Dietrich, Robert Montgomery, Marlon Brando, Veronica Lake; 94 stars in all. Full background on photographers, technical aspects, much more. Total of 160pp. 8⅜ x 11¼.   23352-9 Pa. $6.00

THE NEW YORK STAGE: FAMOUS PRODUCTIONS IN PHOTO-GRAPHS, edited by Stanley Appelbaum. 148 photographs from Museum of City of New York show 142 plays, 1883-1939. *Peter Pan, The Front Page, Dead End, Our Town,* O'Neill, hundreds of actors and actresses, etc. Full indexes. 154pp. 9½ x 10.   23241-7 Pa. $6.00

DIALOGUES CONCERNING TWO NEW SCIENCES, Galileo Galilei. Encompassing 30 years of experiment and thought, these dialogues deal with geometric demonstrations of fracture of solid bodies, cohesion, leverage, speed of light and sound, pendulums, falling bodies, accelerated motion, etc. 300pp. 5⅜ x 8½.   60099-8 Pa. $4.00

THE GREAT OPERA STARS IN HISTORIC PHOTOGRAPHS, edited by James Camner. 343 portraits from the 1850s to the 1940s: Tamburini, Mario, Caliapin, Jeritza, Melchior, Melba, Patti, Pinza, Schipa, Caruso, Farrar, Steber, Gobbi, and many more—270 performers in all. Index. 199pp. 8⅜ x 11¼.   23575-0 Pa. $6.50

J. S. BACH, Albert Schweitzer. Great full-length study of Bach, life, background to music, music, by foremost modern scholar. Ernest Newman translation. 650 musical examples. Total of 928pp. 5⅜ x 8½. (Available in U.S. only)   21631-4, 21632-2 Pa., Two-vol. set $11.00

COMPLETE PIANO SONATAS, Ludwig van Beethoven. All sonatas in the fine Schenker edition, with fingering, analytical material. One of best modern editions. Total of 615pp. 9 x 12. (Available in U.S. only)
                              23134-8, 23135-6 Pa., Two-vol. set $15.00

KEYBOARD MUSIC, J. S. Bach. Bach-Gesellschaft edition. For harpsichord, piano, other keyboard instruments. English Suites, French Suites, Six Partitas, Goldberg Variations, Two-Part Inventions, Three-Part Sinfonias. 312pp. 8⅛ x 11. (Available in U.S. only)   22360-4 Pa. $6.95

FOUR SYMPHONIES IN FULL SCORE, Franz Schubert. Schubert's four most popular symphonies: No. 4 in C Minor ("Tragic"); No. 5 in B-flat Major; No. 8 in B Minor ("Unfinished"); No. 9 in C Major ("Great"). Breitkopf & Hartel edition. Study score. 261pp. 9⅜ x 12¼.
                                                23681-1 Pa. $6.50

THE AUTHENTIC GILBERT & SULLIVAN SONGBOOK, W. S. Gilbert, A. S. Sullivan. Largest selection available; 92 songs, uncut, original keys, in piano rendering approved by Sullivan. Favorites and lesser-known fine numbers. Edited with plot synopses by James Spero. 3 illustrations. 399pp. 9 x 12.   23482-7 Pa. $9.95

PRINCIPLES OF ORCHESTRATION, Nikolay Rimsky-Korsakov. Great classical orchestrator provides fundamentals of tonal resonance, progression of parts, voice and orchestra, tutti effects, much else in major document. 330pp. of musical excerpts. 489pp. 6½ x 9¼. 21266-1 Pa. $7.50

TRISTAN UND ISOLDE, Richard Wagner. Full orchestral score with complete instrumentation. Do not confuse with piano reduction. Commentary by Felix Mottl, great Wagnerian conductor and scholar. Study score. 655pp. 8⅛ x 11. 22915-7 Pa. $13.95

REQUIEM IN FULL SCORE, Giuseppe Verdi. Immensely popular with choral groups and music lovers. Republication of edition published by C. F. Peters, Leipzig, n. d. German frontmaker in English translation. Glossary. Text in Latin. Study score. 204pp. 9⅜ x 12¼. 23682-X Pa. $6.00

COMPLETE CHAMBER MUSIC FOR STRINGS, Felix Mendelssohn. All of Mendelssohn's chamber music: Octet, 2 Quintets, 6 Quartets, and Four Pieces for String Quartet. (Nothing with piano is included). Complete works edition (1874-7). Study score. 283 pp. 9⅜ x 12¼. 23679-X Pa. $7.50

POPULAR SONGS OF NINETEENTH-CENTURY AMERICA, edited by Richard Jackson. 64 most important songs: "Old Oaken Bucket," "Arkansas Traveler," "Yellow Rose of Texas," etc. Authentic original sheet music, full introduction and commentaries. 290pp. 9 x 12. 23270-0 Pa. $7.95

COLLECTED PIANO WORKS, Scott Joplin. Edited by Vera Brodsky Lawrence. Practically all of Joplin's piano works—rags, two-steps, marches, waltzes, etc., 51 works in all. Extensive introduction by Rudi Blesh. Total of 345pp. 9 x 12. 23106-2 Pa. $14.95

BASIC PRINCIPLES OF CLASSICAL BALLET, Agrippina Vaganova. Great Russian theoretician, teacher explains methods for teaching classical ballet; incorporates best from French, Italian, Russian schools. 118 illustrations. 175pp. 5⅜ x 8½. 22036-2 Pa. $2.50

CHINESE CHARACTERS, L. Wieger. Rich analysis of 2300 characters according to traditional systems into primitives. Historical-semantic analysis to phonetics (Classical Mandarin) and radicals. 820pp. 6⅛ x 9¼. 21321-8 Pa. $10.00

EGYPTIAN LANGUAGE: EASY LESSONS IN EGYPTIAN HIERO-GLYPHICS, E. A. Wallis Budge. Foremost Egyptologist offers Egyptian grammar, explanation of hieroglyphics, many reading texts, dictionary of symbols. 246pp. 5 x 7½. (Available in U.S. only) 21394-3 Clothbd. $7.50

AN ETYMOLOGICAL DICTIONARY OF MODERN ENGLISH, Ernest Weekley. Richest, fullest work, by foremost British lexicographer. Detailed word histories. Inexhaustible. Do not confuse this with Concise Etymological Dictionary, which is abridged. Total of 856pp. 6½ x 9¼. 21873-2, 21874-0 Pa., Two-vol. set $12.00

A MAYA GRAMMAR, Alfred M. Tozzer. Practical, useful English-language grammar by the Harvard anthropologist who was one of the three greatest American scholars in the area of Maya culture. Phonetics, grammatical processes, syntax, more. 301pp. 5⅜ x 8½. 23465-7 Pa. $4.00

THE JOURNAL OF HENRY D. THOREAU, edited by Bradford Torrey, F. H. Allen. Complete reprinting of 14 volumes, 1837-61, over two million words; the sourcebooks for *Walden*, etc. Definitive. All original sketches, plus 75 photographs. Introduction by Walter Harding. Total of 1804pp. 8½ x 12¼. 20312-3, 20313-1 Clothbd., Two-vol. set $50.00

CLASSIC GHOST STORIES, Charles Dickens and others. 18 wonderful stories you've wanted to reread: "The Monkey's Paw," "The House and the Brain," "The Upper Berth," "The Signalman," "Dracula's Guest," "The Tapestried Chamber," etc. Dickens, Scott, Mary Shelley, Stoker, etc. 330pp. 5⅜ x 8½. 20735-8 Pa. $4.50

SEVEN SCIENCE FICTION NOVELS, H. G. Wells. Full novels. *First Men in the Moon, Island of Dr. Moreau, War of the Worlds, Food of the Gods, Invisible Man, Time Machine, In the Days of the Comet.* A basic science-fiction library. 1015pp. 5⅜ x 8½. (Available in U.S. only) 20264-X Clothbd. $8.95

ARMADALE, Wilkie Collins. Third great mystery novel by the author of *The Woman in White* and *The Moonstone*. Ingeniously plotted narrative shows an exceptional command of character, incident and mood. Original magazine version with 40 illustrations. 597pp. 5⅜ x 8½. 23429-0 Pa. $6.00

MASTERS OF MYSTERY, H. Douglas Thomson. The first book in English (1931) devoted to history and aesthetics of detective story. Poe, Doyle, LeFanu, Dickens, many others, up to 1930. New introduction and notes by E. F. Bleiler. 288pp. 5⅜ x 8½. (Available in U.S. only) 23606-4 Pa. $4.00

FLATLAND, E. A. Abbott. Science-fiction classic explores life of 2-D being in 3-D world. Read also as introduction to thought about hyperspace. Introduction by Banesh Hoffmann. 16 illustrations. 103pp. 5⅜ x 8½. 20001-9 Pa. $2.00

THREE SUPERNATURAL NOVELS OF THE VICTORIAN PERIOD, edited, with an introduction, by E. F. Bleiler. Reprinted complete and unabridged, three great classics of the supernatural: *The Haunted Hotel* by Wilkie Collins, *The Haunted House at Latchford* by Mrs. J. H. Riddell, and *The Lost Stradivarious* by J. Meade Falkner. 325pp. 5⅜ x 8½. 22571-2 Pa. $4.00

AYESHA: THE RETURN OF "SHE," H. Rider Haggard. Virtuoso sequel featuring the great mythic creation, Ayesha, in an adventure that is fully as good as the first book, *She*. Original magazine version, with 47 original illustrations by Maurice Greiffenhagen. 189pp. 6½ x 9¼. 23649-8 Pa. $3.50

UNCLE SILAS, J. Sheridan LeFanu. Victorian Gothic mystery novel, considered by many best of period, even better than Collins or Dickens. Wonderful psychological terror. Introduction by Frederick Shroyer. 436pp. 5⅜ x 8½. 21715-9 Pa. $6.00

JURGEN, James Branch Cabell. The great erotic fantasy of the 1920's that delighted thousands, shocked thousands more. Full final text, Lane edition with 13 plates by Frank Pape. 346pp. 5⅜ x 8½.
23507-6 Pa. $4.50

THE CLAVERINGS, Anthony Trollope. Major novel, chronicling aspects of British Victorian society, personalities. Reprint of Cornhill serialization, 16 plates by M. Edwards; first reprint of full text. Introduction by Norman Donaldson. 412pp. 5⅜ x 8½. 23464-9 Pa. $5.00

KEPT IN THE DARK, Anthony Trollope. Unusual short novel about Victorian morality and abnormal psychology by the great English author. Probably the first American publication. Frontispiece by Sir John Millais. 92pp. 6½ x 9¼. 23609-9 Pa. $2.50

RALPH THE HEIR, Anthony Trollope. Forgotten tale of illegitimacy, inheritance. Master novel of Trollope's later years. Victorian country estates, clubs, Parliament, fox hunting, world of fully realized characters. Reprint of 1871 edition. 12 illustrations by F. A. Faser. 434pp. of text. 5⅜ x 8½. 23642-0 Pa. $5.00

YEKL and THE IMPORTED BRIDEGROOM AND OTHER STORIES OF THE NEW YORK GHETTO, Abraham Cahan. Film *Hester Street* based on *Yekl* (1896). Novel, other stories among first about Jewish immigrants of N.Y.'s East Side. Highly praised by W. D. Howells—Cahan "a new star of realism." New introduction by Bernard G. Richards. 240pp. 5⅜ x 8½. 22427-9 Pa. $3.50

THE HIGH PLACE, James Branch Cabell. Great fantasy writer's enchanting comedy of disenchantment set in 18th-century France. Considered by some critics to be even better than his famous *Jurgen*. 10 illustrations and numerous vignettes by noted fantasy artist Frank C. Pape. 320pp. 5⅜ x 8½. 23670-6 Pa. $4.00

ALICE'S ADVENTURES UNDER GROUND, Lewis Carroll. Facsimile of ms. Carroll gave Alice Liddell in 1864. Different in many ways from final Alice. Handlettered, illustrated by Carroll. Introduction by Martin Gardner. 128pp. 5⅜ x 8½. 21482-6 Pa. $2.00

FAVORITE ANDREW LANG FAIRY TALE BOOKS IN MANY COLORS, Andrew Lang. The four Lang favorites in a boxed set—the complete *Red, Green, Yellow* and *Blue* Fairy Books. 164 stories; 439 illustrations by Lancelot Speed, Henry Ford and G. P. Jacomb Hood. Total of about 1500pp. 5⅜ x 8½. 23407-X Boxed set, Pa. $14.95

HOUSEHOLD STORIES BY THE BROTHERS GRIMM. All the great Grimm stories: "Rumpelstiltskin," "Snow White," "Hansel and Gretel," etc., with 114 illustrations by Walter Crane. 269pp. 5⅜ x 8½.
21080-4 Pa. $3.50

SLEEPING BEAUTY, illustrated by Arthur Rackham. Perhaps the fullest, most delightful version ever, told by C. S. Evans. Rackham's best work. 49 illustrations. 110pp. 7⅞ x 10¾.
22756-1 Pa. $2.50

AMERICAN FAIRY TALES, L. Frank Baum. Young cowboy lassoes Father Time; dummy in Mr. Floman's department store window comes to life; and 10 other fairy tales. 41 illustrations by N. P. Hall, Harry Kennedy, Ike Morgan, and Ralph Gardner. 209pp. 5⅜ x 8½.
23643-9 Pa. $3.00

THE WONDERFUL WIZARD OF OZ, L. Frank Baum. Facsimile in full color of America's finest children's classic. Introduction by Martin Gardner. 143 illustrations by W. W. Denslow. 267pp. 5⅜ x 8½.
20691-2 Pa. $3.50

THE TALE OF PETER RABBIT, Beatrix Potter. The inimitable Peter's terrifying adventure in Mr. McGregor's garden, with all 27 wonderful, full-color Potter illustrations. 55pp. 4¼ x 5½. (Available in U.S. only)
22827-4 Pa. $1.25

THE STORY OF KING ARTHUR AND HIS KNIGHTS, Howard Pyle. Finest children's version of life of King Arthur. 48 illustrations by Pyle. 131pp. 6⅛ x 9¼.
21445-1 Pa. $4.95

CARUSO'S CARICATURES, Enrico Caruso. Great tenor's remarkable caricatures of self, fellow musicians, composers, others. Toscanini, Puccini, Farrar, etc. Impish, cutting, insightful. 473 illustrations. Preface by M. Sisca. 217pp. 8⅜ x 11¼.
23528-9 Pa. $6.95

PERSONAL NARRATIVE OF A PILGRIMAGE TO ALMADINAH AND MECCAH, Richard Burton. Great travel classic by remarkably colorful personality. Burton, disguised as a Moroccan, visited sacred shrines of Islam, narrowly escaping death. Wonderful observations of Islamic life, customs, personalities. 47 illustrations. Total of 959pp. 5⅜ x 8½.
21217-3, 21218-1 Pa., Two-vol. set $12.00

INCIDENTS OF TRAVEL IN YUCATAN, John L. Stephens. Classic (1843) exploration of jungles of Yucatan, looking for evidences of Maya civilization. Travel adventures, Mexican and Indian culture, etc. Total of 669pp. 5⅜ x 8½.
20926-1, 20927-X Pa., Two-vol. set $7.90

AMERICAN LITERARY AUTOGRAPHS FROM WASHINGTON IRVING TO HENRY JAMES, Herbert Cahoon, et al. Letters, poems, manuscripts of Hawthorne, Thoreau, Twain, Alcott, Whitman, 67 other prominent American authors. Reproductions, full transcripts and commentary. Plus checklist of all American Literary Autographs in The Pierpont Morgan Library. Printed on exceptionally high-quality paper. 136 illustrations. 212pp. 9⅛ x 12¼.
23548-3 Pa. $12.50

AN AUTOBIOGRAPHY, Margaret Sanger. Exciting personal account of hard-fought battle for woman's right to birth control, against prejudice, church, law. Foremost feminist document. 504pp. 5⅜ x 8½.
20470-7 Pa. $5.50

MY BONDAGE AND MY FREEDOM, Frederick Douglass. Born as a slave, Douglass became outspoken force in antislavery movement. The best of Douglass's autobiographies. Graphic description of slave life. Introduction by P. Foner. 464pp. 5⅜ x 8½.   22457-0 Pa. $5.50

LIVING MY LIFE, Emma Goldman. Candid, no holds barred account by foremost American anarchist: her own life, anarchist movement, famous contemporaries, ideas and their impact. Struggles and confrontations in America, plus deportation to U.S.S.R. Shocking inside account of persecution of anarchists under Lenin. 13 plates. Total of 944pp. 5⅜ x 8½.
22543-7, 22544-5 Pa., Two-vol. set $12.00

LETTERS AND NOTES ON THE MANNERS, CUSTOMS AND CONDITIONS OF THE NORTH AMERICAN INDIANS, George Catlin. Classic account of life among Plains Indians: ceremonies, hunt, warfare, etc. Dover edition reproduces for first time all original paintings. 312 plates. 572pp. of text. 6⅛ x 9¼.   22118-0, 22119-9 Pa.. Two-vol. set $12.00

THE MAYA AND THEIR NEIGHBORS, edited by Clarence L. Hay, others. Synoptic view of Maya civilization in broadest sense, together with Northern, Southern neighbors. Integrates much background, valuable detail not elsewhere. Prepared by greatest scholars: Kroeber, Morley, Thompson, Spinden, Vaillant, many others. Sometimes called Tozzer Memorial Volume. 60 illustrations, linguistic map. 634pp. 5⅜ x 8½.
23510-6 Pa. $7.50

HANDBOOK OF THE INDIANS OF CALIFORNIA, A. L. Kroeber. Foremost American anthropologist offers complete ethnographic study of each group. Monumental classic. 459 illustrations, maps. 995pp. 5⅜ x 8½.
23368-5 Pa. $13.00

SHAKTI AND SHAKTA, Arthur Avalon. First book to give clear, cohesive analysis of Shakta doctrine, Shakta ritual and Kundalini Shakti (yoga). Important work by one of world's foremost students of Shaktic and Tantric thought. 732pp. 5⅜ x 8½. (Available in U.S. only)
23645-5 Pa. $7.95

AN INTRODUCTION TO THE STUDY OF THE MAYA HIEROGLYPHS, Syvanus Griswold Morley. Classic study by one of the truly great figures in hieroglyph research. Still the best introduction for the student for reading Maya hieroglyphs. New introduction by J. Eric S. Thompson. 117 illustrations. 284pp. 5⅜ x 8½.   23108-9 Pa. $4.00

A STUDY OF MAYA ART, Herbert J. Spinden. Landmark classic interprets Maya symbolism, estimates styles, covers ceramics, architecture, murals, stone carvings as artforms. Still a basic book in area. New introduction by J. Eric Thompson. Over 750 illustrations. 341pp. 8⅜ x 11¼.
21235-1 Pa. $6.95

GEOMETRY, RELATIVITY AND THE FOURTH DIMENSION, Rudolf Rucker. Exposition of fourth dimension, means of visualization, concepts of relativity as Flatland characters continue adventures. Popular, easily followed yet accurate, profound. 141 illustrations. 133pp. 5⅜ x 8½.
23400-2 Pa. $2.75

THE ORIGIN OF LIFE, A. I. Oparin. Modern classic in biochemistry, the first rigorous examination of possible evolution of life from nitrocarbon compounds. Non-technical, easily followed. Total of 295pp. 5⅜ x 8½.
60213-3 Pa. $4.00

PLANETS, STARS AND GALAXIES, A. E. Fanning. Comprehensive introductory survey: the sun, solar system, stars, galaxies, universe, cosmology; quasars, radio stars, etc. 24pp. of photographs. 189pp. 5⅜ x 8½. (Available in U.S. only)
21680-2 Pa. $3.75

THE THIRTEEN BOOKS OF EUCLID'S ELEMENTS, translated with introduction and commentary by Sir Thomas L. Heath. Definitive edition. Textual and linguistic notes, mathematical analysis, 2500 years of critical commentary. Do not confuse with abridged school editions. Total of 1414pp. 5⅜ x 8½.   60088-2, 60089-0, 60090-4 Pa., Three-vol. set $18.50

Prices subject to change without notice.

Available at your book dealer or write for free catalogue to Dept. GI, Dover Publications, Inc., 180 Varick St., N.Y., N.Y. 10014. Dover publishes more than 175 books each year on science, elementary and advanced mathematics, biology, music, art, literary history, social sciences and other areas.